AF896448

Artificial Intelligence in Customer Service

Jagdish N. Sheth
Varsha Jain
Emmanuel Mogaji • Anupama Ambika
Editors

Artificial Intelligence in Customer Service

The Next Frontier for Personalized Engagement

palgrave
macmillan

Editors
Jagdish N. Sheth
Emory University
Atlanta, GA, USA

Varsha Jain
MICA
Ahmedabad, India

Emmanuel Mogaji
University of Greenwich
London, UK

Anupama Ambika
Institute of Management Technology
Dubai, UAE

ISBN 978-3-031-33897-7 ISBN 978-3-031-33898-4 (eBook)
https://doi.org/10.1007/978-3-031-33898-4

© The Editor(s) (if applicable) and The Author(s), under exclusive licence to Springer Nature Switzerland AG 2023
This work is subject to copyright. All rights are solely and exclusively licensed by the Publisher, whether the whole or part of the material is concerned, specifically the rights of translation, reprinting, reuse of illustrations, recitation, broadcasting, reproduction on microfilms or in any other physical way, and transmission or information storage and retrieval, electronic adaptation, computer software, or by similar or dissimilar methodology now known or hereafter developed.
The use of general descriptive names, registered names, trademarks, service marks, etc. in this publication does not imply, even in the absence of a specific statement, that such names are exempt from the relevant protective laws and regulations and therefore free for general use.
The publisher, the authors, and the editors are safe to assume that the advice and information in this book are believed to be true and accurate at the date of publication. Neither the publisher nor the authors or the editors give a warranty, expressed or implied, with respect to the material contained herein or for any errors or omissions that may have been made. The publisher remains neutral with regard to jurisdictional claims in published maps and institutional affiliations.

This Palgrave Macmillan imprint is published by the registered company Springer Nature Switzerland AG.
The registered company address is: Gewerbestrasse 11, 6330 Cham, Switzerland

Acknowledgment

We want to thank the reviewers for their extensive and constructive feedback. The reviews helped us to scale up the quality of the chapters. We would also like to thank Mr. Parth Salunke, a second-year doctoral scholar at MICA, India, for helping and supporting us in structuring and formatting the content.

Contents

Notes on Contributors

Anupama Ambika is Lecturer in Marketing at the Institute of Management Technology, Dubai, UAE. Her research interests are at the intersection of technology and consumer behavior, and digital marketing. She has published her research in leading marketing and consumer behavior journals. She has over ten years of industry experience with multinational companies, including IBM and Capgemini.

Rijul Chaturvedi is a fellow (Marketing) at the National Institute of Industrial Engineering (NITIE), Mumbai, India.

Yuanyuan Cui is a Ph.D. candidate at Auckland University of Technology and has completed Bachelor of Business (double-majoring in Marketing and Economics) and Bachelor of Business (Honors, specialized in Marketing). Her current role as Research Assistant and Teaching Assistant has allowed her to further develop research skills while also developing valuable creativity skills, analytical reasoning abilities, and interpersonal relationships that she has built on through her academic experience. She is currently pursuing her Ph.D. study (ABD) and a career as an independent researcher where she applies her knowledge and gains valuable experience, contributing to more advanced theory and body of knowledge.

Damini Goyal Gupta PGDM (Marketing), IMT Nagpur, is FPM Scholar at MICA, Ahmedabad, India. Her research interests are at the intersection of digital marketing, advertising, consumer behavior, and luxury branding. Damini has published in marketing journals such as the *Journal of Macromarketing and*

Marketing Intelligence and Planning. Damini has also presented her research works at several international conferences, such as AMA and AMS. Damini is a recipient of several awards for her research accomplishments, marketing campaigns, and activations, including the "Research Award 2021"—first-year doctoral program, MICA, India.

Varsha Jain is the AGK Chair, Professor of Marketing, and co-chairperson of the doctoral program and research at the MICA, India. She is an associate editor (AE) at the *Journal of Advertising Research* (A Journal), the *Journal of Consumer Marketing* (A Journal), and the *Journal of Consumer Behavior* (*JCB*, A Journal). She received the best AE award from JCB in 2022; she is an editorial board member at the International Journal of Information Management (A* Journal) and *Journal of Business Research* (A Journal) and co-track chairperson at, advertising academy of marketing science, USA the annual conference, 2023. She is Vice Chair of International Relations, Higher Education, Special Interest Group, and the American Marketing Association, USA. She has authored over 120+ publications, including papers in A* and A journals. Jain is the recipient of more than 23 international and national awards and gold medals in scholarship. In her research career, she is a visiting guest at Emory Business School, Atlanta, USA, visiting scholar at Greenwich University, London, and The Medill School, Northwestern University, USA. Her interest areas are ABCD—advertising, branding, consumer behavior, and digital (primarily, immersive technologies, including AI, AR, VR, and Meta). She is the coeditor for special issues related to AI, social media influencers, and Metaverse for JCB, which is due by 2023. She has co-authored books with Jag Sheth and Don Schultz, including *Consumer Behavior: A Digital Native*. She is now working on the international edition of a consumer behavior book in the digital context, AI as the next frontier for personalized engagement for services, customer relationship management for digital context with Sheth, and purpose-driven branding and other branding books with David Aaker and another book on qualitative research methods in the digital age, Macmillan, UK.

Sakshi Kathuria is Assistant Professor of Marketing Area at the Fortune Institute of International Business (FIIB), India. Her current research focus spans social media, sustainability, and Industry 4.0 in the context of emerging economies. Her work has been published in leading journals such as *Business Strategy and the Environment, Journal of Public Affairs, Technology in Society*, and *International Journal of Entrepreneurial Behavior & Research*. She is also the Managing Editor of the Emerald Book Series "*Review of Management Literature*."

Werner H. Kunz is Professor of Marketing and Director of the Digital Media Lab at the University of Massachusetts Boston. His work has been published, among others, in the *Journal of Service Research, Journal of Retailing, International Journal of Research in Marketing, Tourism Management, and Computational Statistics* and was awarded multiple times. He is the Senior Editor of the *Journal of Service Research* and a current board member of the Service Research Special Interest Group (SERVSIG) of the American Marketing Association (AMA), the primary professional association of service researchers with over 2000 community members worldwide.

Emmanuel Mogaji has a Ph.D. in Marketing and is Senior Lecturer in Advertising and Marketing Communications at the University of Greenwich, London. Emmanuel's primary areas of interest are ABCDE of Marketing Communications—Advertising | Branding | Communications | Digital | Ethics. He has studied attitudes toward brands and advertisements through various media, such as print and social. He is a fellow of the Higher Education Academy (HEA), demonstrating his teaching philosophy and commitment to encouraging experiential learning. He is the author of *Brand Management: An Introduction Through Storytelling*, published by Palgrave, *Introduction to Advertising: Understanding and Managing the Advertising Process, Marketing in Healthcare-Related Industries*, and *Emotional Appeals in Advertising Banking Services*. He has also co-edited five books on higher education, green and fashion marketing.

Shubhadeep Mukherjee is an assistant professor (lecturer) at the Department of Computer Science, University of Bath, U.K. He holds a visiting researcher position at the Department of Applied Mathematics and Theoretical Physics (DAMTP), University of Cambridge, and works in the area of deep learning and inverse problems in imaging. My research objective is to develop novel, computationally efficient, and provably convergent algorithms for challenging image analysis problems by combining model- and data-driven approaches. Prior to joining Bath, he was a postdoc at the Cambridge Image Analysis (CIA) group, DAMTP, University of Cambridge, and was a member of the AI and image analysis team for the all-in-one cancer imaging project conducted by a group of mathematicians, computer scientists, and radiologists at the University of Cambridge. Before joining Cambridge, he worked as a postdoc in the imaging group at the Department of Mathematics, KTH, Sweden, and completed his Ph.D. in September 2018 at the Indian Institute of Science (IISc.), Bangalore.

Nguyen Phong Nguyen is Head of the Department of Research Administration–International Relations and a lecturer at the School of Accounting, University of

Economics, Ho Chi Minh City. He obtained a Doctor of Business Administration degree at Western Sydney University, Australia (in 2014) and a Master of Business (Accounting) degree at Monash University, Australia (in 2007). He is also a full member of Certified Practising Accountants, Australia. His research focuses on intra-organizational coopetition strategy, the interfaces between accounting and marketing, employees' well-being, and the psychological climate in organizations. He has publications in the *Journal of Business Research, Human Resource Management, Industrial Marketing Management, Business Strategy and the Environment, European Journal of Marketing, Public Management Review, Journal of Product and Brand Management, Journal of Accounting and Public Policy*, and others.

Stefanie Paluch is Full Professor and Chair of Service and Technology Marketing at the School of Business and Economics at RWTH Aachen University. She is a renowned expert in services and technology marketing, focusing on the challenges of digital transformation and the management of innovative technology-based services. Stefanie began her academic journey by studying economics and social sciences at the University of Dortmund and the Ceram Business School in France. She then went on to earn her Ph.D. in 2011 at the T.U. Dortmund University later served as Assistant Professor of Service and Technology Management at the same institution from 2012 to 2014. In 2018, Stefanie was appointed a research fellow at King's Consumer and Organization Data Analytics (CODA) Research Centre at King's Business School in London, and in 2019, she was made a senior fellow at the Centre for Relationship Marketing and Service Management at the Hanken School of Economics in Helsinki. She has received several international research and teaching awards. Her research has been published in leading international academic journals, including the *Journal of Service Research, Journal of Business Research, Journal of Service Management, International Journal of Hospitality Management, Journal of Service Marketing*, and the *Journal of Service Theory and Practice*.

Valentina Pitardi is Assistant Professor of Marketing at Surrey Business School. She holds a Ph.D. and M.Phil. in Marketing and a B.A. in Sociology and Communication Study from the Sapienza University of Rome. Her research interests include the consumer psychology of technology usage, digital consumer engagement, and the role of negative emotions in consumer behavior. Her most recent research focuses on the psychological implications of AI agents' usage. Her works have been published in several international outlets such as the *Journal of Business Research, Psychology and Marketing, Journal of Public Policy* *and Marketing*, and *Journal of Service Management*. She has received various best paper awards for her research and is on the editorial review boards of major marketing and consumer behavior journals. Valentina also serves as Associate Editor of the *International Journal of Market Research*.

Sudhir Rana is a faculty of marketing and strategy at the College of Healthcare Management & Economics at Gulf Medical University, UAE. He holds a Ph.D. degree with the Ministry of Human Resource Development Scholarship from the Government of India and Post-Doctorate from Putra Business School (AACSB Accredited), Universiti of Putra, Malaysia. Rana has been associated with the Fortune Institute of International Business India and the Maastricht School of Management, Kuwait. He has delivered 150+ workshop sessions, keynote speeches, and webinars. His research area is Consumer Behavior and Emerging Markets. His research is published in the *Journal of Business Research*, the *International Journal of Emerging Markets*, and the *Journal of Promotion Management*, among others. He is a well-known scholar and editor in the field and contributed to the progression of the journal *FIIB Business Review* which is the most preferred journal from Sage Publishing and has launched as well as chaired the Doctorate Program at Fortune Institute of International Business, India.

Gourav Roy is an FPM (fellow program in management) scholar at MICA, Ahmedabad (India). His research interests are digital marketing, tourism marketing, and gamification, and he has two years of experience in sales and marketing. He has also interned with Nielsen and IMRB as a research analyst and has kept active roles in qualitative and quantitative research. He has authored two book chapters and a research paper.

Parth Salunke is presently a doctoral scholar at MICA, India, and his research interest lies at the intersection of advertising, branding, and consumer behavior. He has been awarded Research Excellence Award, 2022, MICA, and holds a Master's degree in Management Studies (Marketing) from the University of Mumbai. He has served in several leadership positions during his educational career and worked in the capacity of B.D. & Marketing Executive for three years in the corporate sector. Research piqued his interest during the early stage of his career, and he has participated in several international conferences wherein he received three times "Best Research Paper Award." He is a recipient of more than 70 accolades for his passion for public speaking and debates. He was awarded the Sarvajanik Vidya Sadak Award in 2018 and the Outstanding Aspirant Award in 2020 for his achievements and commendable performance among several graduates.

Jagdish N. Sheth holds the Charles H. Kellstadt Chair Professor of Business at Emory University, Georgia, USA. He has published over 300 research papers and 30 books on consumer behavior, multivariate methods, competitive strategy, relationship marketing, and emerging markets. His book, The Theory of Buyer Behavior (1969) with John A. Howard is a marketing classic. Sheth is a fellow of the American Psychological Association, American Marketing Association (AMA), and the Academy of Marketing Science. He was President of the Association for Consumer Research. Sheth is the recipient of all the top four academic awards bestowed by the AMA.

George Spais is an adjunct professor of the University of Patras, and adjunct faculty member of the Hellenic Open University. He has held adjunct faculty posts at the University of Nicosia (Cyprus), Canterbury Christ Church University, U.K. (Athens Campus), London Metropolitan University, U.K. (Athens Campus), Kentucky University, USA (Athens Campus), New Haven University, USA (Athens Campus), Athens University of Economics and Business (Greece), Agricultural University of Athens (Greece), University of Ioannina (Greece), University of Thessaly (Greece), and at other Greek universities. He is a member of the editorial boards of 12 refereed scholarly journal publications, an Editorial Advisory Board Member of Asia Pacific Journal of Marketing and Logistics, and FIIB Business Review/FBR. He serves as Associate Editor of the Journal of Promotion Management, and he edited special issues of journal publications and book volumes as a guest editor. He is currently a co-guest editor of the special issue of the *Journal of Consumer Behaviour* "A.I. Marketing and AI-Based Promotions Impact on Consumer Behavior and the Avoidance of Consumer Autonomy Threat." He is an elected Fellow Member (FCIM) and certified Chartered Marketer of the Chartered Institute of Marketing (U.K.). He is a trainer and educator for major significant Greek agencies, associations, and unions such as the Greek National Centre for Public Administration and Local Government, the Hellenic Confederation of Commerce and Entrepreneurship, and the Social Multicenter of ADEDY (Supreme Administration of the Greek Civil Servants Trade Unions). He is a member of the Academy of Marketing Science (USA), the European Marketing Academy of the European Institute for Advanced Studies in Management (Belgium), the Marketing Academy (U.K.), the American Marketing Association (USA), the Marketing Educators' Association (USA), the Greek Marketing Academy (Greece), Hellenic Institute of Marketing (Greece), and the Hellenic Adult Education Association.

Patrick van Esch is Assistant Professor of Marketing at the Coles College of Business at Kennesaw State University. His industry experience ranges from sales and marketing, operations, general management, and non-profit board positions across Australia, U.K., and the USA. Patrick has taught business, management, and marketing at universities/colleges in both Australia and the USA, and his research interests include: artificial intelligence, augmented reality, digital marketing, social media marketing, consumer behavior, social marketing, gender/diversity, and LGBTQ advertising.

Sanjeev Verma is presently working as Professor (Marketing) at the National Institute of Industrial Engineering (NITIE), Mumbai, India. He is an active researcher, has published widely in international journals, and authored one book. His papers have been published in various international journals of repute like *Psychology and Marketing*, *Journal of Interactive Marketing*, *Tourism Review*, *Government Information Quarterly*, *Journal of Marketing Communications*, *Journal of Internet Commerce*, *Journal of Modelling Management*, *Journal of Global Marketing*, *International Journal of Marketing and Philanthropy*, *Global Business Review*, etc.

Jochen Wirtz is Vice Dean of MBA Programs and Professor of Marketing at the National University of Singapore (NUS). Further, he is an international fellow of the Service Research Center at Karlstad University, Sweden, an Academic Scholar at the Cornell Institute for Healthy Futures (CIHF) at Cornell University, and a Global Faculty of the Center for Services Leadership (CSL) at Arizona State University. Previously, Wirtz was the founding director of the dual degree UCLA–NUS Executive MBA Program (from 2002 to 2017), an associate fellow at the Saïd Business School, the University of Oxford (from 2008 to 2013), and a founding member of the NUS Teaching Academy, the NUS think-tank on education matters (from 2009 to 2015).Wirtz holds a Ph.D. in services marketing from the London Business School. His research focuses on service marketing and management, and has published over 200 academic articles. His books include the best-selling *Intelligent Automation: Learn How to Harness Artificial Intelligence to Boost Business & Make Our World More Human* (2021), *Services Marketing: People, Technology, Strategy* (9th edition, 2022), and *Essentials of Services Marketing* (4th edition, 2023). With translations and adaptations for over 26 countries and regions, and combined sales of over 1 million copies, they have become globally leading textbooks. In recognition of his excellence in research and teaching, Wirtz has received over 50 awards, including the prestigious Christopher Lovelock Career Contributions to the Services Discipline Award (the highest recognition of the American Marketing Association (AMA) service community), the Academy of Marketing Science (AMS) Outstanding Marketing Teacher Award (the highest recognition of teaching excellence of

AMS globally), and the top university-level Outstanding Educator Award at NUS. Wirtz has been an active management consultant, working with international consulting firms, including Accenture, Arthur D. Little, KPMG, and major service firms worldwide.

Thomas Wittkop has been teaching and researching in the field of "Marketing and Human Resource Management in Civil Engineering" in the Department of Civil Engineering at the Ostwestfalen-Lippe University of Applied Sciences and Arts. The 43-year-old sees apprenticeships as an important link between knowledge transfer and career preparation. His personal concern is to impart sound know-how and important social skills to the students and maintain their interest in teaching. Wittkop completed his economics and social sciences studies at the University of Dortmund in 1999. Subsequently, he worked as a research assistant for six years, and in 2005 he received his doctorate on the topic of "Intercultural competence using the example of German expatriates in China—A qualitative analysis." In his dissertation, he evaluates interviews with representatives of DaimlerChrysler, Henkel, and ThyssenKrupp in Beijing, Shanghai, and Hong Kong to determine which relevant intercultural requirements expatriates must meet. He remained loyal to the Ruhr area when he was employed from 2005 to 2006 as assistant to the management at Hauhinco Maschinenfabrik GmbH & Co K.G. in Sprockhövel. Another essential station in his career is his professorship for marketing at the SRH University of Applied Sciences in Hamm, which he held from 2007 to 2009. Wittkop is a nature lover who passionately discovers the world and its different cultures on his travels. "I enjoy contact with foreign cultures because I always want to relativize my views." His partnership gives him a lot of strength for upcoming tasks: "Without them, I would never have achieved what I have achieved."

List of Figures

List of Tables

1

Artificial Intelligence in Customer Service: An Introduction to the Next Frontier to Personalized Engagement

Varsha Jain, Jagdish N. Sheth, Emmanuel Mogaji, and Anupama Ambika

Introduction

Artificial intelligence is playing an integral role in enhancing business operations (Tarafdar et al., 2019; Abdulquadri et al., 2021). Organizations are now expected to evaluate whether they can automate their business operations (Khalil & Abdelli, 2022). With a focus on customer services, this book reiterated the vast prospects of AI as the next frontier to personalized engagement. Huge data are emerging from different consumer

V. Jain (✉)
MICA, Ahmedabad, India
e-mail: varsha.jain@micamail.in

J. N. Sheth
Emory University, Atlanta, CA, USA

E. Mogaji
University of Greenwich, London, UK
e-mail: E.O.Mogaji@greenwich.ac.uk

© The Author(s), under exclusive license to Springer Nature Switzerland AG 2023

J. N. Sheth et al. (eds.), *Artificial Intelligence in Customer Service*,
https://doi.org/10.1007/978-3-031-33898-4_1

touchpoints with brands, there are increasing computing powers, and consumers are growing with huge expectations from their brands. There is now no excuse for brands to ignore the prospects of AI in achieving personalized engagement and increasing customer satisfaction (Mogaji et al., 2020). However, there is still a limited theoretical and practical understanding of managing this digital transformation (Dwivedi et al., 2021).

Specifically, this book sets out to meet this need and offer the much-needed theoretical and practical understanding of how artificial intelligence (AI) can enable customer service functions to accomplish higher customer engagement, superior experience, and increased levels of well-being. Today, customer expectations of 24/7 availability and personalization, self-help trends, and market pressures for lower cost & higher efficiency have led to the acceptance of AI as a vital component of customer service (Mogaji & Nguyen, 2022; Kunz & Wirtz, 2023). While it is prudent to deploy AI for superior customer service, businesses may learn to choose the right tool and optimal integration and strike the appropriate equilibrium between AI and humans for exceptional benefits (Abushammala et al., 2022; Chatterjee et al., 2019). Many AI projects have failed to take off or succeed due to the lack of attention to the above-mentioned components. Hence, it is essential to understand AI's recent advancements and applications in customer service through real-world examples, as intended by this book.

The relevance of artificial intelligence (AI) has expanded significantly in every sphere (Owoc et al., 2021; Sharma et al., 2022). Applying AI in the customer service arena can substantially improve the customer service experience. In the present scenario, AI works best when integrated into current real-world applications to enhance the experiences. Though today most businesses are exploring the implementation of AI, there are several misconceptions about this technology's possibilities and limitations.

A. Ambika
MICA, Ahmedabad, India

Institute of Management Technology, Dubai, UAE
e-mail: anupama@imt.ac.ae

Hence, developing resources to educate prospective practitioners and the academic community about AI's theoretical and practical aspects in customer service is highly significant. Therefore, this book intends to provide the frameworks, best practices, key components, future opportunities, and possible challenges to guide practitioners toward implementing effective AI-enabled customer service.

Outline of the Book

Chapter 2 by Kunz and Wirtz (2023) provided an overview of AI in customer services, recognizing how AI shapes the service revolution and enhances brands' ability to engage with customers. This chapter draws on their previous studies on corporate digital responsibility in service firms and their ecosystems (Wirtz et al., 2022), cost-effective service excellence, service robots, and artificial intelligence (Wirtz, 2020), and Intelligent Automated and Service Robots (Wirtz et al., 2021). The chapter reiterated how AI-based customer services often operate within larger virtual networks that allow them to access external and internal data. This enables them to learn from their environments and subsequently decide without human input. The chapter concludes with practical implications for managers, suggesting that organizations should incorporate corporate digital responsibility (CDR) and create a set of shared values, norms, and actionable guidelines for the responsible use of technology throughout the complete cycle (Wirtz et al., 2022).

Chaturvedi and Verma (2023) focused on the Opportunities & Challenges of AI-Driven Customer Service in Chap. 3 of this book. The authors reiterated AI's immense potential and opportunities to acquire a competitive advantage but are open to challenges. Deployment of AI technologies for customer service embeds dichotomic dilemmas known as conundrums and paradoxes, such as the Personalization-Privacy Paradox, the Uncanny Valley Paradox, and so on. On the bright side, factors such as personalization, convenience, and anthropomorphism unpack value co-creation. On the dark side, technology anxiety, privacy concern, lack of human interaction, and loss of control slip value co-destruction. A customer-centric service ecosystem requires the balanced use of AI-driven value co-creation while mitigating value-co-destruction concerns. The chapter concludes with a strategic framework to overcome challenges with balancing strategies to yield superior AI-enabled customer service.

AI in Customer Service Strategy was discussed by Cui and van Esch (2023), recognizing that companies across industries are embracing artificial intelligence (AI) to create more frictionless and memorable customer service experiences. However, successfully implementing AI customer service strategies requires companies to realize the value of strategically deploying AI and human agents to optimize customer service. To realize the full potential of AI, companies ideally ought to foster a sustainable AI-human symbiosis. In this technology-oriented ecosystem, AI and humans assist, complement, and learn from each other to offer a seamless service experience for customers. After outlining the significance of an AI customer service strategy, this chapter highlights how companies could capitalize on AI as a key element in successfully implementing customer service strategies. Specifically, they outlined the conceptualization of AI, followed by a brief overview of AI customer service strategies. They also provided key steps to successfully implement an AI customer service strategy—including triage, prospect, pre-consumption, and post-consumption management—and showcased relevant real-world examples.

Pitardi (2023) shared their thoughts on Personalized and Contextual AI Services Experience in Chap. 5. The chapter acknowledges how AI service applications are becoming increasingly pervasive in our everyday life. At the same time, companies are adopting technology like chatbots, robots, and conversational assistants to offer customer services and support in various settings. Yet, it still needs to be determined which are the conditions under which AI-based services may improve the overall customer experience. This chapter offers a three-way framework to interpret and evaluate the AI-based service customer experience. It identifies relevant contextual, personal, and technological characteristics that can affect the AI-service delivery. The proposed framework allows marketers to identify the best fit between the context, the technology configuration, and customer characteristics, providing companies with practical and actionable suggestions to enhance the customer experience.

Chapter 6 by Spais (2023) explored how brands can engage Self Service for better customer experience. Based on consumer ethics, morality, social response theory, and the crucial role of consumers' social reaction types (the "Social Customer Relationship Management view"), this chapter presents an exciting perspective of artificial intelligence (AI) by proposing a model for planning the development of a self-service automation system in retailing. The intention is to improve the AI experience with the deontological moral framework as the fundamental basis of the system based on a priori and a posteriori consumer knowledge before launching the enabling technologies. The "critical realistic view" of the two examples is vital in this discussion. Amazon's cashier-less Go Grocery retail locations and the adoption of fully autonomous self-checkout systems ("Grab and Go technology") is the first critical example. The second critical example is the Computer Are Social Actor paradigm with consumers on the Taobao app, the merchant's customer service chatbot agent, and the perceived communication quality. The chapter summarizes some exciting implications for industry, academia, and society.

Understanding the Role of Data in Artificial Intelligence-based Personalization was the focus of Chap. 7 by Mukerjee (2023). The author noted that the arrival of big data capabilities has fundamentally transformed how organizations function, making data ubiquitous and inimitable for decision-making. Although this transformation has created massive opportunities for organizations to engage customers, personalize recommendations and generate more revenue, this has also led to the increased complexity of data (structured, unstructured, and semi-structured), resulting in newer challenges in utilizing it for decision-making. Manually collecting and analyzing such vast and varied data is nearly impossible. The end-to-end data life cycle consists of several interrelated processes: data collection, cleaning, exploration, analysis, and implementation. This chapter discusses a conceptual framework to describe how data is fundamentally changing the business landscape and the methods through which data could be collected, analyzed, and reported to ensure maximum customer engagement and personalization.

Nguyen and Mogaji (2023) discuss the huge prospects of Artificial Intelligence for Seamless Experience across Channels in Chap. 8. This chapter analyses the use of AI for seamless experiences across channels. Implementing seamless experiences across different channels is necessary, given the increased utilization of different retail channels. Furthermore, with the growing amount of data and computer processing capabilities, it is essential to explore how businesses can integrate artificial intelligence (AI) and machine learning (ML) with the data to enhance their business operations. The chapter entails a literature review to enumerate key findings later utilized to devise managerial implications and policy recommendations. Significant managerial implications include the need for business leaders to invest in AI-enabled value chains, evaluate the deployment of AI in the business, and make necessary enhancements by collaborating with AI experts to create better customer service programs provided in the chapter. Likewise, the chapter offers policy recommendations that revolve around implementing policies that support consumer data protection, skills development, and talent acquisition in AI.

Chapter 9 by Gourav Roy et al. (2023) evaluated data processing and AI-technology integration for personalized services. With artificial intelligence-aided technology, service organizations enable better personalization of services. These technologies offer optimal solutions by understanding customer choices, needs, and preferences by collecting various customer data points. Here, to provide personalization, the customer data is utilized to predict the needs and requirements, wherein data processing and AI-technology integration undertake the key role. Thus, this chapter aims to present a holistic view of how data is processed, enabled by artificial intelligence to provide personalized services. The chapter also highlights that artificial intelligence is a facilitator of personalization, and the customer is the final decision maker. However, the process is a cyclical loop wherein the customer data points are collected, and data gets refined at each stage to offer better-personalized services.

Gupta and Jain (2023) provided some theoretical insights in Chap. 10 by discussing AI with Ethics and Privacy for Personalized Customer Services. The authors noted that the increase in the use of AI by service providers helps serve customers 24/7/365. Furthermore, AI helps build a flexible system that educates and empowers consumers while facilitating service providers in gathering detailed insights and streamlining communication. However, using AI in customer support services questions ethics and privacy. Ethics relates to using fair and inclusive methods of collecting, storing, and disseminating customer data. The chapter provides implications to academicians and industrialists to help them understand the part of using AI with ethics and privacy in enhancing the customer experience with customer support service. The chapter concludes by suggesting areas for future research further to improve customer experiences through AI in customer services.

Chapter 11 by Kathuria and Rana (2023) provided a roadmap for linking customer e-service quality with Artificial Intelligence-Based Business Environment. The digital technology-enabled retail system consists of various cutting-edge digital technologies, and consumer shops improve customer service quality and trust in digital technologies. The fast-changing consumer preferences develop a competitive environment for the product and services because consumer e-satisfaction/services (well-being) are essential for developing business strategies to improve customer technological well-being. Developing a linkage from human social interaction, the concept of consumer trust provides a valid reason for measuring and identifying the relationship between humans and AI-driven technological transformation. However, many consumer studies do not discuss AI-driven service quality parameters. Therefore, this chapter presents various customer service quality expectations and their relevance to the AI-enabled business environment. The chapter provides future research scope for future research to extend the proposed linkages for greater consumer e-services. The recommendation of the present study is to help product and service market users understand the importance of customer service quality in AI-driven business environments.

Chapter 12 is the last chapter of the book. It investigates the role of risk on employees' willingness to collaborate with artificial intelligence and its impact on well-being in the workplace. In this chapter, Stefanie Paluch and Wittkop (2023) noted that collaborating with Artificial-Intelligence-based services (AIBS or AI-based services) will be inevitable due to a superior job performing abilities and increasing computational capabilities for employees in the service context. AI-based services in the form of virtual (e.g., chatbot) or physical (e.g., service robot) embodiments, which are characterized by a higher degree of autonomy and are intended to augment the tasks of human employees in a highly collaborative way, will increasingly become an integral part of employee's work environment. Semi-structured interviews and a focus group discussion have been used to identify constructs driving employees' perception of AI-based services and their impact on well-being. Combined with a thorough literature analysis, relations between the identified constructs have been derived from introducing a conceptual model of employees' willingness to collaborate (WTC). This study identifies technology characteristics, job demands, resources, and control as constructs, directly influencing employees' perceived risk of AI-based services.

Conclusion

We had the pleasure of working with 21 contributing authors worldwide, both from practice and academia, with different experiences and expertise in AI, customer services, robotics, brand management, and other forms of business and management. We have endeavored to focus on using AI for effective customer service and personalized engagement and experience to present the strategies for developing a competitive advantage through AI-enabled customer service. This offers insights on enhancing the well-being of customers and employees through AI-enabled customer services, elaborating on the opportunities and challenges of using AI for effective customer services. Finally, it explains the importance of data, processes, and technology integration to ensure effective customer service with the help of AI.

It is imperative to recognize that we have not been able to cover the whole spectrum of AI in customer service and personalized engagement; there are still many more grounds to fill. We have made a significant contribution to this stream of research. With many of the chapters clearly identifying areas for future research and providing future research questions, we anticipate this book will spur further research interest in AI and customer services. Importantly, we hope to make an impact and improve business operations through these contributions and the practical insights provided for managers and policymakers.

References

Abdulquadri, A., Kieu, T. A., & Nguyen, N. P. (2021). Digital transformation in financial services provision: A Nigerian perspective to the adoption of chatbot. *Journal of Enterprising Communities: People and Places in the Global Economy, 15*(2), 258–281.

Abushammala, S. N., Abushawish, H. F., Jalambo, M. O., & Maghari, A. Y. (2022). Transforming into smart business: The role of artificial intelligence. In *Future of organizations and work after the 4th industrial revolution: The role of artificial intelligence, big data, automation, and robotics* (pp. 81–102). Springer International Publishing.

Chatterjee, S., Ghosh, S. K., Chaudhuri, R., & Nguyen, B. (2019). Are CRM systems ready for AI integration? A conceptual framework of organizational readiness for effective AI-CRM integration. *The Bottom Line, 32*(2), 144–157.

Chaturvedi, R., & Verma, S. (2023). Opportunities & challenges of AI-driven customer service. In J. Sheth, V. Jain, E. Mogaji, & A. Ambika (Eds.), *Artificial intelligence in customer service: Next frontier to personalized engagement* (pp. 11–20). Palgrave.

Cui, Y., & van Esch, P. (2023). AI in customer service strategy. In J. Sheth, V. Jain, E. Mogaji, & A. Ambika (Eds.), *Artificial intelligence in customer service: Next frontier to personalized engagement* (pp. 11–20). Palgrave.

Dwivedi, Y. K., Hughes, L., Ismagilova, E., Aarts, G., Coombs, C., Crick, T., et al. (2021). Artificial Intelligence (AI): Multidisciplinary perspectives on emerging challenges, opportunities, and agenda for research, practice and policy. *International Journal of Information Management, 57*, 101994.

Gourav Roy, G., Jain, V., & Salunke, P. (2023). Data processing and AI-technology integration for personalized services. In J. Sheth, V. Jain, E. Mogaji, & A. Ambika (Eds.), *Artificial intelligence in customer service: Next frontier to personalized engagement* (pp. 11–20). Palgrave.

Gupta, G., & Jain, V. (2023). Use of AI with ethics and privacy for personalized customer services. In J. Sheth, V. Jain, E. Mogaji, & A. Ambika (Eds.), *Artificial intelligence in customer service: Next frontier to personalized engagement* (pp. 11–20). Palgrave.

Kathuria, S., & Rana, S. (2023). Maintaining customer trust and e-service quality in an artificial intelligence-based business environment. In J. Sheth, V. Jain, E. Mogaji, & A. Ambika (Eds.), *Artificial intelligence in customer service: Next frontier to personalized engagement* (pp. 11–20). Palgrave.

Khalil, A., & Abdelli, M. E. A. (2022). Do digital technologies influence the relationship between the COVID-19 crisis and SMEs' resilience in developing countries? *Journal of Open Innovation: Technology, Market, and Complexity, 8*(2), 100–109.

Kunz, W., & Wirtz, J. (2023). AI in customer services – A service revolution in the making. In J. Sheth, V. Jain, E. Mogaji, & A. Ambika (Eds.), *Artificial intelligence in customer service: Next frontier to personalized engagement* (pp. 11–20). Palgrave.

Mogaji, E., & Nguyen, P. N. (2022). Managers' understanding of artificial intelligence in relation to marketing financial services: Insights from a cross-country study. *International Journal of Bank Marketing, 40*(6), 1272–1298.

Mogaji, E., Soetan, T., & Kieu, T. (2020). The implications of artificial intelligence on the digital marketing of financial services to vulnerable customers. *Australasian Marketing Journal., 29*(3), 235–242.

Mukerjee, S. (2023). The role of big data. In J. Sheth, V. Jain, E. Mogaji, & A. Ambika (Eds.), *Artificial intelligence in customer service: Next frontier to personalized engagement* (pp. 11–20). Palgrave.

Nguyen, P. N., & Mogaji, E. (2023). Artificial intelligence for seamless experience across the channels. In J. Sheth, V. Jain, E. Mogaji, & A. Ambika (Eds.), *Artificial intelligence in customer service: Next frontier to personalized engagement* (pp. 11–20). Palgrave.

Owoc, M. L., Sawicka, A., & Weichbroth, P. (2021, August). Artificial intelligence technologies in education: benefits, challenges and strategies of implementation. In *Artificial Intelligence for Knowledge Management: 7th IFIP WG 12.6 International Workshop, AI4KM 2019, Held at IJCAI 2019, Macao, China, August 11, 2019, revised selected papers* (pp. 37–58). Springer International Publishing.

Pitardi, V. (2023). Personalised and contextual AI- services experience. In J. Sheth, V. Jain, E. Mogaji, & A. Ambika (Eds.), *Artificial intelligence in customer service: Next frontier to personalized engagement* (pp. 11–20). Palgrave.

Sharma, H., Soetan, T., Farinloye, T., Mogaji, E., & Noite, M. D. F. (2022). AI adoption in universities in emerging economies: Prospects, challenges and recommendations. In *Re-imagining educational futures in developing countries: Lessons from Global Health crises* (pp. 159–174). Springer International Publishing.

Spais, G. (2023). Engaging self-service. In J. Sheth, V. Jain, E. Mogaji, & A. Ambika (Eds.), *Artificial intelligence in customer service: Next frontier to personalized engagement* (pp. 11–20). Palgrave.

Stefanie Paluch, S., & Wittkop, T. (2023). Investigating the role of risk on employees' willingness to collaborate with artificial intelligence and its impact on wellbeing in the workplace. In J. Sheth, V. Jain, E. Mogaji, & A. Ambika (Eds.), *Artificial intelligence in customer service: Next frontier to personalized engagement* (pp. 11–20). Palgrave.

Tarafdar, M., Beath, C. M., & Ross, J. W. (2019). Using AI to enhance business operations. *MIT Sloan Management Review, 60*(4), 6–23.

Wirtz, J. (2020). Strategic pathways to cost-effective service excellence. In *The Routledge Handbook of Service Research Insights and Ideas* (pp. 257–279). Routledge.

Wirtz, J., Kunz, W., & Paluch, S. (2021). The service revolution, intelligent automation and service robots. *European Business Review, 29*(5), 909.

Wirtz, J., Tarbit, J. J., Hartley, N., & Kunz, W. (2022). Corporate digital responsibility: Dealing with ethical, privacy and fairness challenges of AI. *Journal of AI, Robotics & Workplace Automation, 1*(4), 325–328.

Theme I

Introduction to AI and Customer Services

2

AI in Customer Service: A Service Revolution in the Making

Werner H. Kunz and Jochen Wirtz

Introduction

Welcome to the Digital Service Revolution

The industrial revolutions of the late eighteenth century boosted our economy by introducing automated manufacturing processes. This revolution paved the way for higher quality, cost-effective products available to mass markets and freed people from tedious and laborious manual work—drastically improving our standard of living! With the proliferation of technology and its quick integration into service organizations, a digital revolution in services is just beginning—much like the industrial revolution brought sweeping changes to manufacturing. As cost savings

W. H. Kunz (✉)
University of Massachusetts, Boston, MA, USA
e-mail: werner.kunz@umb.edu

J. Wirtz
National University of Singapore, Singapore, Singapore
e-mail: jochen@nus.edu.sg

© The Author(s), under exclusive license to Springer Nature Switzerland AG 2023
J. N. Sheth et al. (eds.), *Artificial Intelligence in Customer Service*,
https://doi.org/10.1007/978-3-031-33898-4_2

will compete away in market economies, this digital service transformation can drastically improve our quality of life by mechanizing and streamlining services like banking, insurance, logistics, healthcare, and education (Bock et al., 2020; Huang & Rust, 2018; Wirtz et al., 2018).

Artificial intelligence (AI), digitization, and AI-based customer service technologies such as service robots are driving a revolution in the service sector. These technologies are improving the customer experience, service quality, and productivity (Wirtz et al., 2018), and hardware and software are becoming more advanced, powerful, and cost-effective. Examples of hardware include physical service robots, smart self-service technologies, autonomous vehicles, drones, and their components like processors, sensors, and cameras. The software includes code, analytics systems, image processing, speech processing, biometrics, virtual and augmented reality, geo-tagging, mobile technologies, cloud technologies, low-code platforms, robotic process automation, machine learning, and deep learning (Bornet et al., 2021).

Huang and Rust see AI as machines that exhibit aspects of Human Intelligence (HI) characterized by self-learning and connectivity (Huang & Rust, 2021). Kaplan and Haenlein describe AI as "a system's ability to interpret external data correctly, to learn from such data, and to use those learnings to achieve specific goals and tasks through flexible adaptation" (Kapplan & Haenlein, 2019, p. 17). In the context of this article, we follow Rai and colleagues and refer to AI as "the ability of a machine to perform cognitive functions that we associate with human minds, such as perceiving, reasoning, learning, interacting with the environment, problem-solving, decision-making, and even demonstrating creativity" (Rai et al., 2019, p. 3).

AI-based customer services will revolutionize the service industry, leading to unparalleled innovation that can drastically enhance customers' experiences, improve the quality of service, and significantly boost productivity (Wirtz & Zeithaml, 2018). Due to the constant growth of robot technology and AI, service robots have become prevalent. Service robots are "system-based autonomous and adaptable interfaces that interact, communicate and deliver service to an organization's customers" (Wirtz et al., 2018, p. 909). These machines can function independently, needing little to no human input for task completion (Joerling et al.,

2019), execute tasks effectively by utilizing the provided service script as well as prior information (Huang & Rust, 2018), and are widely acknowledged to be a critical driver of innovation (Rust & Huang, 2014).

AI-based customer services often operate within larger virtual networks that allow them to access external and internal data. This enables them to learn from their environments and subsequently decide without human input. For example, service robots equipped with cameras and sensors can use facial or voice recognition to identify customers and provide personalized services based on customer profiles accessed through interconnected systems.

In the following discussion, we will explore the benefits and challenges of AI-based customer services, highlighting their strengths and considering their implications for the future of customer service.

Benefits of AI-Based Customer Service

By implementing AI-based customer service offerings, companies can offer users direct benefits in various ways. One benefit is that the service is informed and designed using algorithms that constantly learn from historical and real-time data. This allows it to adapt and personalize the service experience to fit the customer's needs. This results in a more relevant and effective service that addresses the customer's wants and needs (Wixom & Ross, 2017). Services like Amazon Alexa or Google Assistant use previous search and engagement history to learn about a user's preferences and evolve the experience with each interaction.

AI-based services provide greater convenience for consumers as they are faster at handling service requests than humans and are available 24/7 (Newell & Marabelli, 2015). In addition, the low incremental cost for a service request allows consumers to use them frequently without incurring significant expenses. An example of this increased efficiency and productivity can be seen in live translation apps that use data-driven technology to translate a conversation in real-time. These apps allow users to have synchronous conversations in a foreign language without needing a human service representative, saving them time and effort in searching for solutions. They also provide users with instant answers to their questions rather than waiting for a representative to assist them.

From the user's perspective, AI-based services are often very cost-effective as they can be provided at a fraction of the cost of high-touch human-delivered services and, in some cases, are offered entirely for free (Gu et al., 2018). An example is AI-based proofreading services, which can accurately proofread documents at a low cost due to the computer-generated results. In addition, the low or zero financial cost to the user allows for new user experiences created and delivered by algorithms. For instance, social media platforms like Snapchat, TikTok, and Instagram use smart filters powered by machine learning and deep learning algorithms to augment video feeds with additional features like beautification or virtual accessories. These experiences would not be possible without the advanced capabilities of AI.

Further, AI-powered customer service is significantly less expensive than traditional human-delivered service and can be easily scaled with minimal additional cost. This includes virtual service robots, such as chatbots, and "visible" ones, like holograms. For example, an airport could use hologram-based humanoid service robots to assist passengers and answer common questions, such as providing information about arrivals and departures and directing travelers to the correct check-in counters and airport hotels. These holograms require only low-cost hardware, such as a camera, microphone, speaker, and projector. In addition, they do not take up physical space, allowing, for example, travelers to easily pass through them as needed while pushing their luggage carts (Wirtz et al., 2021).

The COVID-19 pandemic has increased the use of medical service robots to care for patients with contagious illnesses. One example of an AI-based social robot used for this purpose is Ari. These robots, often AI-powered, can interact with patients to alleviate feelings of isolation, dispense medicine, monitor vital signs remotely, and disinfect hospitals. They can also help ensure that patients and visitors follow necessary regulations and maintain safe distances from one another (Schoepfer & Etemad-Sajadi, 2020).

AI-based services can also empower customers to do more for themselves in collaboration with an algorithm (Viswanathan et al., 2017). This can give them a sense of control and privacy, as no other human service personnel is involved. In addition, this type of service can lead to a deeper level of engagement from the customer in the service process. An example is online dating services that use computational profile matching to identify potential partners.

Incorporating data, analytical insights, or actions into service experiences benefits companies and customers (Wixom et al., 2020). For companies, using data and analytics in service experiences can improve efficiency, productivity, and access to novel insights into customer preferences. This allows companies to provide optimal, flexible interactions in real time. In addition, personalizing and customizing services on a mass scale can also create significant barriers to switching, providing a sustained competitive advantage. Because of these benefits, investments in data and analytics technology and talent, such as data scientists, have been a top priority for executives for the past decade (Kappelman & Sinha, 2021; Mogaji & Nguyen, 2022).

Challenges of AI-Based Customer Service

While the digital service revolution offers many exciting possibilities, it also brings potential downsides and risks for service users (Belk, 2021; Breidbach & Maglio, 2020; Koshiyama et al., 2021; Puntoni et al., 2021). The ease of capturing and analyzing customer data presents challenges beyond what George Orwell could have imagined in his classic novel *Nineteen Eighty-Four*. There are various security breach risks, such as data leaks and hacking. Even intentionally or unintentionally, service providers risk misusing customer data (Beke et al., 2018). For example, the Match group and its subsidiary—Tinder and Grindr, entered into controversy for misusing customer data. This online dating conglomerate has violated EU privacy laws by disclosing their customer's information, such as sexual orientation, HIV status, and GPS locations, with other firms without informing the customers.

In addition to data and privacy concerns, ethical dilemmas are posed by digital technology advancements such as AI, intelligent automation, and machine learning (Belk, 2021; Breidbach & Maglio, 2020). These technologies include unparalleled power and pose a question about their usage and impact on society (Brynjolfsson & Mcafee, 2017; Bornet et al., 2021). For instance, using data to create the indicators and variables such as creditworthiness scores or safe driving practices and then feeding them into AI systems for decision-making raises ethical questions about this automated process. This can lead to unfair consequences such as higher insurance premiums or interest rates. Further, specific demographics or groups, particularly marginalized sections such as women, a certain ethnicity, and elderly women, may face potential discrimination. These affect the individuals negatively (Gillis & Spiess, 2019).

Digital services can also have negative consequences, such as addictive consumption behaviors, including algorithmic reinforcement suppressing self-esteem and triggering emotional dependency on dating apps (Courtois & Timmermans, 2018; Her & Timmermans, 2020), impulsive purchasing patterns that facilitate obesity (Montgomery et al., 2019), inhibition of self-regulation behaviors in online gaming and 'loot box'-type (i.e., in-game) purchases (Brooks & Clark, 2019; Li et al., 2019), and conditioning reward-response behaviors toward emotional validation on social platforms (Kuss & Griffiths, 2017; Montag et al., 2019). At a higher level, service robots and AI cause concerns related to dehumanization, social isolation, loss of autonomy and dignity, social engineering, and more (Belk, 2021; Čaić et al., 2018; Leidner & Tona, 2021; Vandemeulebroucke et al., 2018). Even if introduced with the best intentions, malleable AI systems can be at risk of exploitation for unintended purposes (Koshiyama et al., 2021).

Corporate Digital Responsibility (CDR) as a Response to AI-Related Challenges

Service organizations need to recognize that their customers can use AI systems and digital technologies in harmful ways and may have unintended and unwanted consequences for society as a whole (Lobschat et al., 2021). This was expressed nicely by Wade (2020), "The risks to humanity of poor or unethical digital practices are increasing rapidly and can no longer be ignored. … corporate entities need to join the debate." AI poses particular challenges for service firms because they tend to have closer relationships with customers, more touchpoints, and process more consumer data than goods companies. Services are also easier to digitize than physical goods, leading to an expanding data stream that can be integrated into service value chains. AI is often used to process and analyze these data and train models to make predictions and optimize customer experiences.

Unfortunately, research that guides organizations related to digital technologies in the face of ethical dilemmas is scant, especially in a service context (For a deeper excuse, refer to Wirtz et al., 2022). In this context, Wirtz et al. (2022) suggest the concept of Corporate Digital Responsibility (CDR) as an answer to the new AI-based challenges of service companies and define CDR "as the principles underpinning a service firm's ethical, fair, and protective use of data and technology when engaging with customers within their digital service ecosystem" (p. 1). CDR is particularly important for service markets because the potential ethical concerns based on the disruptive force of digital technology are exceptionally high (Bock et al., 2020).

Organizational goals can sometimes conflict with the adoption of good CDR practices. These tensions can be viewed as trade-offs that service companies must navigate to be both profitable and ethical or as a "utilitarian perspective weighing the benefits and costs" (Hermann, 2022, p. 44). For example, the privacy literature has developed the "privacy calculus" to address such trade-offs from the consumer perspective, suggesting that consumers weigh the benefits and costs of their privacy decisions (Gouthier et al., 2022; Smith et al., 2011). Similarly, Wirtz et al. (2022) argue that service firms weigh the benefits and costs of good CDR to determine their level of engagement in CDR. They illustrate these tensions through a CDR calculus shown in Fig. 2.1.

No service company is completely immune to CDR-questionable behaviors and decision-making. Still, the service organization of the future needs to maintain a reputation of being trustworthy regarding CDR. Using the CDR calculus and considering service companies' CDR behaviors as trade-offs between benefits and costs can provide critical insights beyond the current focus on risks and concerns. Therefore,

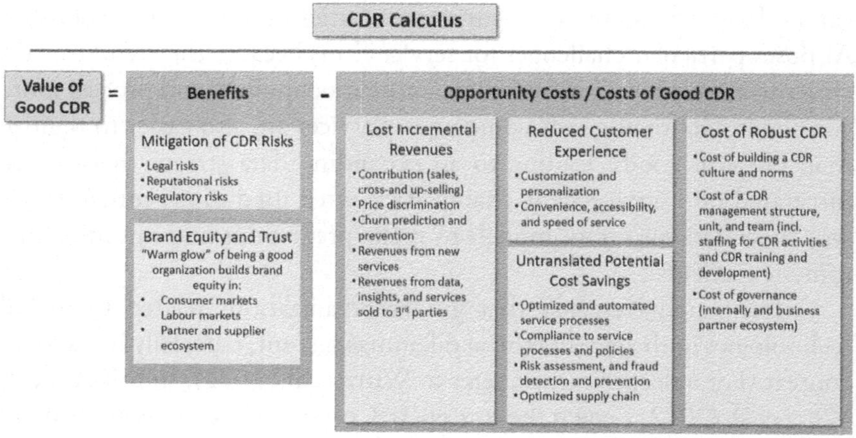

Fig. 2.1 A service firm's CDR calculus. Note: Adapted from Wirtz, Jochen, Werner Kunz, Nicole Hartley, and James Tarbit (2022), "Corporate Digital Responsibility in Service Firms and Their Ecosystems," *Journal of Service Research*, published online first

service organizations should assess their CDR calculus concerning their data practices and technologies. Wirtz et al. (2022) expect service firms to engage in good CDR if the calculus is positive or not too negative, meaning that firms are willing to forgo some profit to avoid "doing bad." However, if the forgone profits become too high, good CDR behaviors become less likely, and regulation may be necessary. Floridi (2021) shares a somewhat pessimistic view and concludes that the era of self-regulation to address ethical challenges related to AI is over. While self-regulation has advantages (e.g., it is faster and more agile and promotes an ethically constructive dialogue that is essential as digital technology develops rapidly), it may not always be sufficient to address these ethical challenges. He concluded that "the time has come to acknowledge that, much as it was worth trying, self-regulation did not work. … Self-regulation needs to be replaced by the law; the sooner, the better" (Floridi, 2021, p. 622).

The above discussion shows clearly that although the usage of AI-based systems brings up ethical challenges, the benefits of AI-based services are also strong. Thus, based on the insights from the CDR calculus, companies will likely strive to deploy more AI-based systems than less (hopefully while keeping CDR principles in mind). Given this likely scenario, it is worth investigating deeper how AI-based service systems differ from traditional service delivery.

AI-Based Customer Service vs. Traditional Service Delivery

AI-based customer services like chatbots, physical service robots, and conversational agents differ from traditional service delivery in various ways. In the following, we discuss some of the most general differences.

Emotional Touch vs. Customized Tech

Frontline employees are often considered the face of a service company, as the quality of the service is determined by their skills, training, emotions, personality, and attitude. The personal touch provided by human employees can be a key differentiator for service organizations, depending on the company's strategy. Personal service involves genuine emotions being exchanged between human beings. In contrast, AI-based customer interfaces do not feel or express real emotions. This is significant because the literature on service management makes a distinction between "deep acting," in which employees express their genuine feelings, and "surface acting," in which employees exhibit phony, superficial, emotional responses (e.g., Wirtz & Jerger, 2017). The emotional expressions of AI will, for the foreseeable future, be false and not based on real feelings. Customers will likely realize this and react to artificial intelligence (AI)-displayed emotions differently from genuine, honest emotions from human frontline personnel (Wirtz et al., 2018). However, because AI-based customer service agents are less likely to experience emotional fatigue than human employees, they may be better able to simulate surface-acted emotions (similar to automated social presence) (van Doorn et al., 2017).

Duffy (2003) asserts that anthropomorphic characteristics in the shape or behavior of the robot are necessary for meaningful social contact between a human and a robot. Strong anthropomorphic traits can cause people to have unduly optimistic expectations of a robot's capabilities, which can leave them disappointed. In other words, a person expects a robot to behave more like an actual human, the more realistic its appearance. For this reason, Duffy (2003) proposes that a "synthetic human" should not be the ideal social robot.

Individual Person vs. System-Based Approach

There is a distinction between a human and a robot. Humans as employees represent heterogeneity, such as varying personalities, biases, perceptions, and skills observed over time and among all employees. These frontline employees require training to learn a routine and memorize information for efficient usage of various IT systems. This process is not seamless and takes time. However, AI-based customer services are connected with knowledge databases when acting in the frontline service system. They use available information from Customer Relationship Management (CRM) systems and external sources such as social media and the Internet to provide services.

High vs. Low Incremental Cost

Human workers cannot scale. Each individual drives up the expense of the business. In contrast, the enormous economies of scale and scope of AI-based customer services mostly cover the expenditures associated with research and development. For instance, even though virtual roots are much less expensive than increasing the workforce, physical robots nevertheless incur incremental costs. Virtual robots, in contrast, are expected to be used at very low incremental costs (Wirtz et al., 2018). Table 2.1 summarizes additional noteworthy variations.

Table 2.1 clearly shows the difference in the service delivery for human versus AI-based customer services. Based on our previous illustrations of AI-based services, we will discuss the following key implications that the digital service revolution has for service organizations.

Table 2.1 Contrasting frontline employees with AI-based customer services

Dimension	Service Employees	AI-Based Customer Service
Training and learning	• Act as individuals, individual learning • Need training • Limited memory and access	• Act as part of systems, are connected, system learning • Upgradable, system-wide • Virtually endless memory and access
Customer experience	• Heterogeneous output • Customization and personalization depend on employee skill and effort • Unintended biases • Have genuine emotions • Can engage in deep acting • Can engage in out-of-box thinking and creative problem solving	• Homogenous output • Customization and personalization can be delivered on the scale at consistent quality and performance • Potentially no biases • Can mimic emotions • Can engage in surface acting • Limited out-of-box thinking, has rule-bound limits
Firm strategy	• Service employees can be a source of competitive advantage • High incremental cost • Low economies of scale and scope • Differentiation in service can be based on better hiring, selection, training, motivation, and organization of service employees	• AI-based customer services are unlikely to be a source of competitive advantage as service robot solutions are likely to be supplied by third-party providers (very much like ATMs are sold to banks) • Low incremental cost • High economies of scale and scope • Economies of scale and scope and related network and service platform effects will become important sources of competitive advantage

Adapted from Jochen Wirtz, Paul Patterson, Werner Kunz, Thorsten Gruber, Vinh Nhat Lu, Stefanie Paluch, and Antje Martins (2018), "Brave New World: Service Robots in the Frontline", *Journal of Service Management*, Vol. 29, No. 5, p. 909, https://doi.org/10.1108/JOSM-04-2018-0119.

Implications for Service Organizations

This upheaval revolution in the service industry will have far-reaching effects on firms. The following are some of the most pressing concerns facing service organizations:

The Service Sector Is at an Inflection Point

Similar to the start of the Industrial Revolution, when manufacturing underwent significant advancements, we are at a critical juncture in the development of the service sector. Companies that can adopt new technology, launch, or update their services, and alter their business models will benefit greatly from this change. Traditional service providers should also make use of AI infusion and digital transformation to boost productivity and quality while reinvigorating their brand image through cutting-edge marketing techniques. They will be able to maintain their competitiveness over time if they adopt these changes in their business model.

Reconstructing the Organizational Frontline

Organizations will need to be transformed and reorganized to implement AI-based customer service. Strong leadership, backing, and the ability and willingness of the workforce to adapt are all necessary for this transformation. RPA, fundamental programming, analytics, and technical troubleshooting are just a few of the new abilities that employees may be given, as well as new tasks and responsibilities. This implies that human service representatives' abilities and competencies will need to change in the future, and this transformation will impact the labor market. In order for the digitalization and AI spirit to be internalized and become part of the company's culture at all levels, not just at the client interface, the organization needs to be ready for this shift.

More Human-Robot Collaboration

We don't think AI-powered service robots will entirely replace customer support workers today or in the future. We disagree with this perspective. Robots will probably be able to do cognitive and analytical activities of previously unheard-of complexity in the following decades. However, human beings will continue to provide most services that are essentially emotional or social and call for genuine feelings (see Fig. 2.2).

As a result, robots will likely take the place of people in most routine or conventional tasks but not in all service scenarios. We continue to think that creating strong customer relations is essentially the responsibility of those working in human services. The fundamental building blocks of trust are their sincere feelings and compassionate, charitable behavior, which are now impossible for robots to reproduce.

In the future, we believe that collaborative hybrid human-robot teams will be the preferred method of service delivery for many services. By

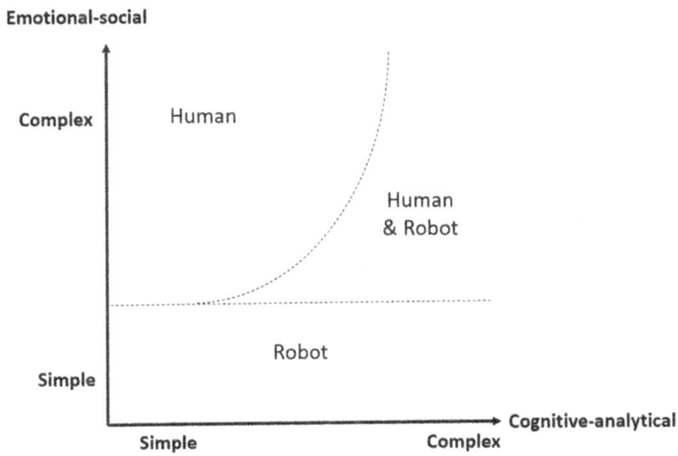

Fig. 2.2 Service delivery is based on the complexity of emotional and cognitive tasks. Adapted from Jochen Wirtz, Paul Patterson, Werner Kunz, Thorsten Gruber, Vinh Nhat Lu, Stefanie Paluch, and Antje Martins (2018), "Brave New World: Service Robots in the Frontline," *Journal of Service Management*, Vol. 29, No. 5, p. 909, https://doi.org/10.1108/JOSM-04-2018-0119

fusing the benefits of AI and human service agents, these hybrid teams will increase the company's efficiency and level of service.

An Opportunity for Cost-effective Service Excellence

We believe that collaborative hybrid teams of humans and robots will be the future service paradigm for many more intricate service settings. By fusing the benefits of AI with human workers, these hybrid teams will increase the company's productivity and level of service. Additionally, robots are particularly well-suited for providing highly customized services due to their extensive knowledge base, access to data, including client profiles in CRM systems, and processing capacity. In order to provide unmatched client connection, businesses should concentrate on deploying co-creation teams made up of robots, staff, and customers and controlling and fine-tuning their use.

Although AI-based customer service is not a panacea for all service problems, it can be a beneficial tool to raise the standard of customer care. Service personnel might enter the encounter and offer help through their emotional or social abilities when situations become more complicated or call for individual attention or recovery. The overall quality of the service will probably improve as a result of these novel ways of interacting.

Mitigating Potential Deployment Risks

To address potential prejudice, misconceptions, and anxieties of AI-based customer services, such as the perceived loss of the human touch, algorithm aversion, and consumer privacy. Organizations should incorporate corporate digital responsibility (CDR) and create a set of shared values, norms, and actionable guidelines for the responsible use of technology throughout the complete cycle (Wirtz et al., 2022). This may entail thinking about issues like data collection (e.g., using biometrics or social media accounts), data use (e.g., creating variables like a healthiness index or financial score), data decision-making (approving loans and determining interest rates), and data retirement (e.g., determining when data on a bounced payment is removed from the company's database).

In conclusion, AI-based customer services have the potential to revolutionize the service industry and deliver extraordinary gains in productivity, service quality, and customer experience all at once. In other words, as the industrial revolution did for manufactured goods, the service revolution could significantly raise our standard of living.

References

Beke, F. T., Eggers, F., & Verhoef, P. C. (2018). Consumer informational privacy: Current knowledge and research directions. *Foundations and Trends in Marketing, 1*(11), 1–71.

Belk, R. (2021). Ethical issues in service robotics and artificial intelligence. *The Service Industries Journal, 41*(13–14), 860–876.

Bock, D. E., Wolter, J. S., & Ferrell, O. C. (2020). Artificial intelligence: Disrupting what we know about services. *Journal of Services Marketing, 34*(3), 317–334.

Bornet, P., Barkin, I., & Wirtz, J. (2021). *Intelligent automation: Welcome to the world of hyperautomation*. World Scientific Books.

Breidbach, C. F., & Maglio, P. (2020). Accountable algorithms? The ethical implications of data-driven business models. *Journal of Service Management, 31*(2), 163–185.

Brooks, A. G., & Clark, L. (2019). Associations between loot box use, problematic gaming and gambling, and gambling-related cognitions. *Addictive Behaviors, 96*, 26–34.

Brynjolfsson, E., & Mcafee, A. (2017). Artificial intelligence, for real. *Harvard Business Review*. https://starlab-alliance.com/wp-content/uploads/2017/09/AI-Article.pdf

Čaić, M., Odekerken-Schröder, G., & Mahr, D. (2018). Service robots: Value co-creation and co-destruction in elderly care networks. *Journal of Service Management, 29*(2), 178–205.

Courtois, C., & Timmermans, E. (2018). Cracking the tinder code: An experience sampling approach to the dynamics and impact of platform governing algorithms. *Journal of Computer-Mediated Communication, 23*(1), 1–16.

Duffy, B. R. (2003). Anthropomorphism and the social robot Author links open overlay panel. *Robotics and Autonomous Systems, 42*(3–4), 177–190

Floridi, L. (2021). The end of an era: From self-regulation to hard law for the digital industry. *Philosophy & Technology, 34*(4), 619–622.

Gillis, T. B., & Spiess, J. L. (2019). Big data and discrimination. *The University of Chicago Law Review, 86*(2), 459–488.

Gouthier, M. H. J., Nennstiel, C., Kern, N., & Wendel, L. (2022). The more the better? Data disclosure between the conflicting priorities of privacy concerns, Information Sensitivity & Personalisation in E-commerce. *Journal of Business Research, 148*, 174–189.

Gu, X., Kannan, P. K., & Ma, L. (2018). Selling the Premium in Freemium. *Journal of Marketing, 82*(6), 10–27.

Her, Y.-C., & Timmermans, E. (2020). Tinder blue, mental flu? Exploring the associations between tinder use and well-being. *Information, Communication & Society, 24*(9), 1303–1319. https://doi.org/10.1080/1369118X.2020.1764606

Hermann, E. (2022). Leveraging artificial intelligence in Marketing for Social Good—An Ethical Perspective. *Journal of Business Ethics, 179*, 43–61.

Huang, M.-H., & Rust, R. T. (2018). Artificial intelligence in service. *Journal of Service Research, 21*(2), 155–172.

Huang, M.-H., & Rust, R. T. (2021). Engaged to a robot? The role of AI in service. *Journal of Service Research, 24*(1), 30–41.

Joerling, M., Boehm, R., & Paluch, S. (2019). Service robots: Drivers of perceived responsibility for service outcomes. *Journal of Service Research, 22*, 404. https://doi.org/10.1177/1094670519842334

Kappelman, A. C., & Sinha, A. K. (2021). Optimal control in dynamic food supply chains using big data. *Computers & Operations Research, 126*, 105–117.

Kapplan, A., & Haenlein, M. (2019). Siri, Siri, in my hand: Who's the fairest in the land? On the interpretations, illustrations, and implications of artificial intelligence. *Business Horizons, 62*(1), 15–25.

Koshiyama, A., Kazim, E., Treleaven, P., Rai, P., Szpruch, L., Pavey, G., Ahamat, G., Leutner, F., Goebel, R., & Knight, A. (2021). Towards algorithm auditing: A survey on managing legal, ethical and technological risks of AI, ML and associated Algorithms. Working Paper. https://doi.org/10.2139/ssrn.3778998.

Kuss, J. D., & Griffiths, D. M. (2017). Social networking sites and addiction: Ten lessons learned. *International Journal of Environmental Research and Public Health, 14*(3), 311.

Leidner, E. D., & Tona, O. (2021). The CARE theory of dignity amid personal data digitalization. *MISQ, 45*, 343–370.

Li, W., Mills, D., & Nower, L. (2019). The relationship of loot box purchases to problem video gaming and problem gambling. *Addictive Behaviors, 97*, 27–34.

Lobschat, L., Mueller, B., Eggers, F., Brandimarte, L., Diefenbach, S., Kroschke, M., & Wirtz, J. (2021). Corporate digital responsibility. *Journal of Business Research, 122*, 875–888.

Mogaji, E., & Nguyen, P. N. (2022). Managers' understanding of artificial intelligence in relation to marketing financial services: Insights from a cross-country study. *International Journal of Bank Marketing, 40*(6), 1272–1298.

Montag, C., Lachmann, B., Herrlich, M., & Zweig, K. (2019). Addictive features of social media/messenger platforms and freemium games against the background of psychological and economic theories. *International Journal of Environmental Research and Public Health, 16*(14), 1–16.

Montgomery, K., Chester, J., Nixon, L., Levy, L., & Dorfman, L. (2019). Big data and the transformation of food and beverage marketing: Undermining efforts to reduce obesity? *Critical Public Health, 29*(1), 110–117.

Newell, S., & Marabelli, M. (2015). Strategic opportunities (and challenges) of algorithmic decision-making: A call for action on the long-term societal effects of 'Datification'. *The Journal of Strategic Information Systems, 24*(1), 3–14.

Puntoni, S., Reczek, R. W., Giesler, M., & Botti, S. (2021). Consumers and artificial intelligence: An experiential perspective. *Journal of Marketing, 85*(1), 131–151.

Rai, A., Constantinides, P., & Sarker, S. (2019). Next-generation digital platforms: Toward human–AI hybrids. *MIS Quarterly, 43*(1), iii–ix.

Rust, R. T., & Huang, M.-H. (2014). The service revolution and the transformation of marketing science. *Marketing Science, 33*(2), 206–221.

Schoepfer, T., & Etemad-Sajadi, R. (2020). The clever use of robots during COVID-19. Retrieved February 2021 at https://hospitalityinsights.ehl.edu/robots-during-covid-19

Smith, H. J., Dinev, T., & Heng, X. (2011). Information privacy research: An interdisciplinary review. *MIS Quarterly, 35*(4), 989–1015.

van Doorn, J., Mende, M., Noble, S. M., Hulland, J., Ostrom, A. L., Grewal, D., & Petersen, J. A. (2017). Domo arigato Mr. Roboto: Emergence of automated social presence in Organisational frontlines and Customers' Service experiences. *Journal of Service Research, 20*(1), 43–58.

Vandemeulebroucke, T., de Casterlé, D. B., & Gastmans, C. (2018). The use of care robots in aged care: A systematic review of argument-based ethics literature. *Archives of Gerontology and Geriatrics, 74*, 15–25.

Viswanathan, V., Hollebeek, L. D., Malthouse, E. C., Ewa Maslowska, S., Kim, J., & Xie, W. (2017). The dynamics of consumer engagement with Mobile technologies. *Service Science, 9*(1), 36–49.

Wade, M. (2020, April 28). Corporate responsibility in the digital era. *MIT Sloan Management Review.* Accessed January 26, 2021, from (Available at https://sloanreview.mit.edu/article/corporate-responsibility-in-the-digital-era/).

Wirtz, J., & Jerger, C. (2017). Managing service employees: Literature review, expert opinions, and research directions. *Service Industries Journal, 36*(15–16), 757–788.

Wirtz, J., Kunz, W., Hartley, N., & Tarbit, J. (2022). Corporate digital responsibility in service firms and their ecosystems. *Journal of Service Research, 26*, 173.

Wirtz, J., Kunz, W., & Paluch, S. (2021, January–February). *The service revolution, intelligent automation and service robots* (pp. 38–44). European Business Review.

Wirtz, J., Patterson, P. G., Kunz, W. H., Gruber, T., Vinh Nhat, L., Paluch, S., & Martins, A. (2018). Brave New World: Service robots in the frontline. *Journal of Service Management, 29*(5), 907–931.

Wirtz, J., & Zeithaml, V. (2018). Cost-effective service excellence. *Journal of the Academy of Marketing Science, 46*(1), 59–80.

Wixom, B. H., & Ross, J. W. (2017). How to monetize your data. *MIT Sloan Management Review, 58*(3), 1–13.

Wixom, B. H., Someh, I. A., Zutavern, A., & Beath, C. M. (2020). Explanation: A new Enterprise data monetization capability for AI. MIT CISR, 1–14.

3

Opportunities and Challenges of AI-Driven Customer Service

Rijul Chaturvedi and Sanjeev Verma

Introduction

AI-enabled technologies are transforming businesses and customer service (Abdulquadri et al., 2021). Mckinsey analytics report (Chui et al., 2021) highlights AI as the enabler for customization and operational efficiency. A survey of 33 AI-used cases revealed a spurt in revenue with AI adoption in business (Cam et al., 2019). In the global economy, cumulative GDP can rise to 16% by 2030 due to the imbibition of AI in business processes (Bughin et al., 2018). Extant literature establishes the pivotal role of AI and AI-enabled technologies in advancing business. It is the time for researchers and practitioners to mull over the opportunities and challenges of AI in industrial revamping globally (Gerbert et al., 2017). Prescriptive analytics suggests that AI will soon take the contradiction stage in customer service, and personalization-based conversational agents (Siri and Alexa) will instrumentalize customer satisfaction

R. Chaturvedi (✉) • S. Verma
NITIE, Mumbai, India
e-mail: rijul.chaturvedi.2020@nitie.ac.in; sanjeev@nitie.ac.in

© The Author(s), under exclusive license to Springer Nature Switzerland AG 2023

J. N. Sheth et al. (eds.), *Artificial Intelligence in Customer Service*,
https://doi.org/10.1007/978-3-031-33898-4_3

(Lim et al., 2022). The present chapter introduces AI-Driven customer service using five major emerging technologies: Recommender Systems, Conversational Agents, Service Robots, the Internet of Things (IoT), and Extended Reality. Businesses deploy AI-based technologies to achieve competitive advantage through enhanced customer service (Verma et al., 2021; Khalil & Abdelli, 2022). To achieve seamless integration of AI in business, firms must understand and balance the AI-driven customer service opportunities and challenges (Chaturvedi & Verma, 2022; Mogaji et al., 2020). AI embed both bright and dark sides to unfold dichotomic dilemmas popularly known as paradoxes or conundrums of AI. Figure 3.1, with the help of a Venn diagram, represents the interrelation between three sets; the dark side, paradoxes, and the bright sides of AI services. The set of positive factors consists of more elements than the negative set, as indicated by the size of the circles in the diagram. Set representing paradoxes include factors from both sides that interplay simultaneously to create confusion in the mind of the consumers—for example, privacy concern and personalization interplay to unfold the Personalization-Privacy Paradox.

On the brighter side, deploying AI-based technologies into customer service unpacks value co-creation through personalization, on-demand service, security, speed, and convenience (Payne et al., 2021). Conversely,

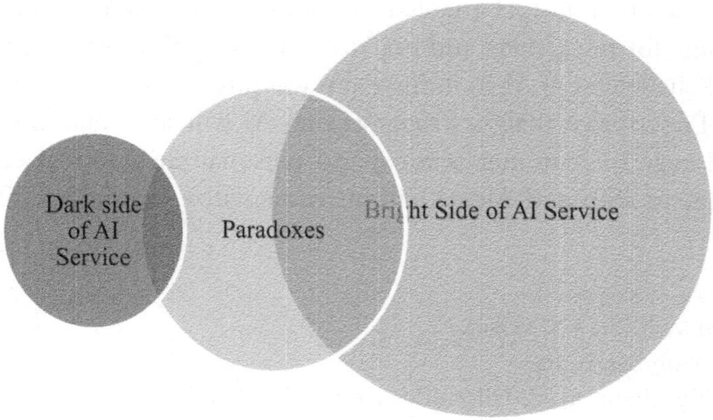

Fig. 3.1 Bright and dark sides of AI

it can co-destruct value through technology anxiety, privacy concern, and psychological barriers (Lalicic & Weismayer, 2021).

The fourth industrial revolution spikes an urgent need to balance the customer-centric service ecosystem driven by AI. Organizations can leverage the benefits of value co-creation while mitigating the factors responsible for value co-destruction. Customer journey interwoven with AI-enabled technologies translates each moment of truth into customer-centric satisfaction. Customer service spindles through value co-creation and co-destruction (Luo et al., 2019; Mogaji & Nguyen, 2022). The present chapter discerns the opportunities and challenges of AI-driven customer service. The proposed strategic framework offers to balance strategies to reduce the effect of value co-destruction factors. The chapter revolves around the following four sections: Customer service in the age of Artificial Intelligence, Conundrums of AI for customer service, Determinants of value co-creation and value co-destruction, and overcoming the challenges.

Customer Service in the Age of Artificial Intelligence

According to Xu et al. (2020), AI-enabled customer services is a technology-enabled market intelligence system and marketing offering. Service quality influences the consumers' behavioral intention, and customer service offers a competitive advantage (Zeithaml et al., 1996). Service quality remained a core marketing strategy for decades. However, knowing how businesses transformed customer services with AI is interesting. Huang and Rust (2018) specified four kinds of intelligence required for customer service- mechanical, analytical, intuitive, and empathetic. AI performs lower-level tasks and higher-order intelligence, such as intuitive and empathetic. Service robots replace mechanical jobs, and analytical tools substitute problem diagnosis technicians. AI Quill can write like human authors, and conversational agents like Replika can provide emotional support (Huang & Rust, 2018). Artificial intelligence for customer service reduces human employment costs and brings operational efficiency to the system. AI-enabled technologies shape customer services in various sectors such as hospitality and tourism, food delivery, e-commerce, entertainment, etc. Some AI features enable personalization, speed, convenience, better technological interaction, and innovation to provide a seamless customer experience. Businesses recognize the role of emerging technologies in bringing new value to their customers and operational efficiency (Rangaswamy et al., 2020). For the transformation of AI-driven customer service, enablers such as machine learning, data analytics, natural language processing, deep learning, neural networks, etc., are essential for value creation. The deployment of AI enablers offers an interactive interface to achieve the end goal of AI customer service (Ameen et al., 2021), customer engagement (Perez-Vega et al., 2021), and customer loyalty (Yoon et al., 2013). The current chapter proposes four layers of objectives to achieve AI-driven Customer service presented as a pyramid (Fig. 3.2).

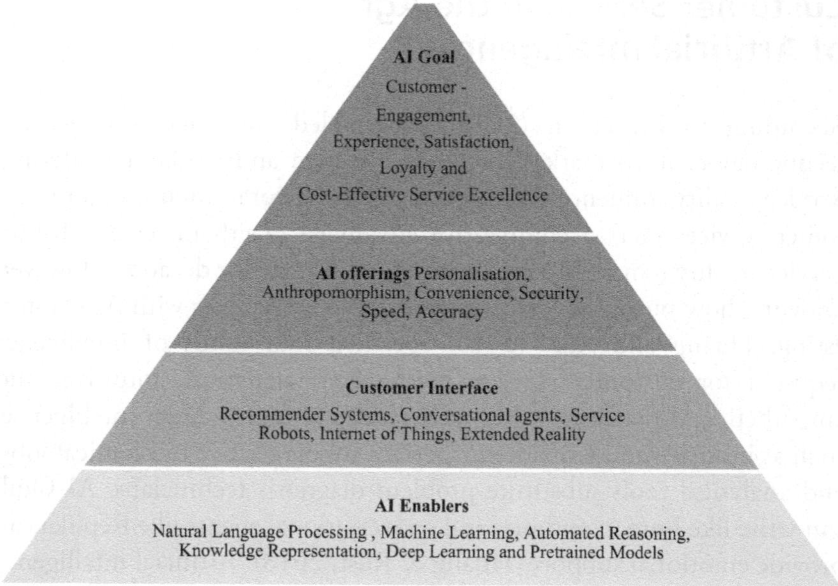

Fig. 3.2 AI-driven customer service pyramid

AI Enablers

Businesses are transforming into AI-driven firms with the deployment of Artificial Intelligence enablers for operational efficiency and customer interface. The following are the key enablers of Artificial intelligence:

Automated Reasoning: Automated reasoning entail programmed algorithms to assist logical problem-solving (Wos et al., 1984) in multiple domains like Tarskian geometry and proof simplification (Beeson et al., 2022). Automated reasoning applications drive various industrial applications like mathematical research, computational logic, robotics, expert systems, etc. (Wos et al., 1985). Recent studies highlight the importance of automated reasoning in neural networks (Leofante et al., 2018), network path estimation (Philp et al., 2020), and functional behavior in robotic applications (Miyazawa et al., 2019).

Knowledge Representation: Artificial intelligence embeds knowledge representation and intelligent systems to solve complex tasks (Newell et al., 1959). Richens (1956) introduced the knowledge graph as a concept of the semantic net. According to Ji et al. (2021), a knowledge graph is "a structured representation of facts consisting of entities, relationships, and semantic descriptions." Knowledge representation unfolds applications in the domain of artificial intelligence, such as computer-based medical consultations (Shortliffe, 2012), the Internet of Things (Wang et al., 2013), and industrial robotic systems (Balakirsky et al., 2012).

Machine Learning: Samuel (1967) defined machine learning as the study of developing learning capability in computers without any explicit programming. It can be classified into subfields based on output: supervised, semi-supervised, unsupervised learning, reinforcement, evolutionary, and ensemble learning (Alzubi et al., 2018). Machine learning has applications in many domains, including mathematics, agriculture, chemistry, cognitive modeling, computer programming, education, expert systems, game playing, image recognition, medical diagnosis, robotics, natural language processing, and speech recognition (Carbonell, 2014).

Natural Language Processing (NLP): According to Liddy (2001), NLP is defined as "A theoretically motivated range of computational techniques for analyzing and representing naturally occurring texts at one or more levels of linguistic analysis to achieve human-like language processing for a range of tasks or applications." NLP has applications for developing chatbots, voice bots, conversational agents, talking robots, intelligent agents, etc.

Deep Learning (DL): DL is a subset of machine learning techniques that allow computational models to run multiple levels of data representation with the help of numerous processing layers. Applications of DL include designing recommendations, visual recognition, and object detection systems (LeCun et al., 2015). DL helps marketers analyze vast amounts of unsupervised data to draw valuable insights about the customers (Najafabadi et al., 2015).

Pretrained Models (PTM): The advancement in machine learning techniques required a massive manual effort to shape the data that can perform desired AI tasks which consume a lot of human energy, time, and monetary expense. However, the best ways to deal with the data-hungry challenge are hidden in the transfer learning technique leading to pretrained models. The approach hires previously learned models and processes them in a fine-tuning stage to solve new problems (Han et al., 2021). PTM is utilized to develop NLP models, image recognition models, visual computing, etc.

Data Processing Technologies (DPT): The internet and connected things generate large volumes, variety, and velocity of data from numerous sources daily. Thus, unlike traditional techniques, emerging data processing technologies are concerned with "Big Data," which includes extensive growing sets of structured, unstructured, and semi-structured data. To process such an amount of data, firms deploy the Hadoop ecosystem. This consists of two main components: file distribution system and MapReduce framework. Parallel clusters and distributed file systems support Hadoop in processing large data sets. In addition, several other modules can be installed in the ecosystem according to the requirement of the target application (Oussous et al., 2018). DPT applications lie in data analysis through the Internet of Things, Cloud Computing, Smart Grid Systems, etc.

AI-Enabled Customer Interface

Over two decades, significant research focused on consumer adoption of new customer interface technologies starting from self-service technologies such as internet shopping, interactive voice response systems, automated payouts, package tracking, and ATM services (Burke, 2002). Emerging AI-powered technologies such as the Internet of Things (IoT), augmented reality/virtual reality/mixed reality (AR/VR/MR) or extended reality (XR), virtual assistants/chatbots/robots are dramatically transforming touchpoints in the customer journey (Hoyer et al., 2020). Businesses are deploying AI-enabled customer interfaces to enhance the service experience of consumers. Some worth-noticing AI-enabled customer interfaces are recommender systems, conversational agents, service robots, extended reality encompassing AR/VR, and the Internet of Things (Table 3.1).

Recommender Systems

Algorithmic systems generate automatic predictions based on users' historical data. Users' "browsing history" hints at the liking and assimilated preferences to recommend (Hsieh et al., 2008). The first recommender system, "Tapestry," recommended documents drawn from newsgroups based on "collaborative filtering" (Goldberg et al., 1992). Collaborative filtering systems collect user ratings and identify the similarities in the rating pattern to recommend a product/service (Melville & Sindhwani, 2010). New algorithms besides collaborative filtering recently added new dimensions to the recommender system. New techniques like content and hybrid filtering models are advanced and flexible. In contemporary times, recommender systems reduce the information overload problem that emerged from the large amount of data available on the Internet (Schafer et al., 1999). Recommender systems applications include OTT platforms like NETFLIX (Lü et al., 2012), shopping websites like Amazon, music applications with personalized recommendations, YouTube, Movie recommendations on various movie collection platforms, and google search for images/documents/books (Park et al., 2012).

Table 3.1 AI-enabled customer interfaces

S. No.	Technology	Benefits
1	Recommender system	• Saves time and effort • Creates a personalized interface • Reduces information overload • Helps in decision making
2	Conversational agents	• Talk like humans • Assist consumers in performing digital tasks • Recommend products that suit customer profile • Provide social presence, companionship, and emotional support
3	Service robots	• Act like humans • Can perform tasks that require mechanical, intuitive, and empathetic skills • Can handle unpleasant and dangerous tasks • Provide social presence, companionship, and emotional support to humans
4	Internet of Things (IoT)	• Auto-monitoring of products • Auto- control functions and features of products • Self-optimization of product performance • Products becoming intelligent, interactive, and playful
5	Extended reality (XR)	• Consumers can try products virtually before buying them • Provide entertainment and gamification effect while purchasing products • Enhanced consumer engagement and a higher knowledge retention • Convenient, easy, and an impressive visual experience

Sophisticated recommender systems enable marketers to personalize interactions with the consumer, reducing information overload problems and assisting consumers in making the right decisions (Aldrich, 2011).

Conversational Agents

Since the rise of the digital era and the emergence of e-commerce, new-age technologies equipped with artificial intelligence have disrupted the way, companies deliver customer service (Buhalis et al., 2019). Today, businesses deploy voice assistants and smart speakers to engage customers with brands (Smith, 2020). These interacting voice assistants or chatbots which communicate like humans are called conversational agents (Laranjo et al., 2018). Conversational agents use natural language processing to respond to customer queries and imitate salespersons to recommend products to consumers (Lucente, 2000). These agents can play three roles in online commerce, as suggested by McGoldrick et al. (2008); first, as assistants in managing various digital tasks such as setting the alarm, information search on the Internet, playing a song, etc. Second, recommending products matching the consumer's requirement, and third, as a social companion for building emotional connections with the consumers, strengthening the customer relationship with the firm. The use of conversational agents is proliferating in multiple service industries such as education (Kerly et al., 2006; Kerly et al., 2007), health (Liu & Sundar, 2018), tourism (Lv et al., 2021), banking (Letheren & Dootson, 2017), etc. Some popular conversational agents are Amazon's Alexa, Apple's Siri, Microsoft's Cortana, and Google's google assistant (Lim et al., 2022). Some digital agents like Replika are specially designed to build companionship with the users and provide emotional support to escape loneliness or for fun/enjoyment (Skjuve et al., 2021).

Service Robots

According to Wirtz et al. (2018), service robots are "system-based autonomous and adaptable interfaces that interact, communicate, and deliver service to an organization's customers." With the rise of advanced digital technologies equipped with AI, robots today are not just mechanical arms performing repetitive tasks but becoming smarter and rapidly replacing human employees in organizations (Huang & Rust, 2018). Market Research Future (2019) indicates that the intelligent robots market was valued at 4.5 billion US dollars in 2017. Surprisingly, this value is expected to reach 15 billion US dollars by 2023. Robots can perform mundane tasks and handle unpleasant/ dangerous jobs, making them more significant (Chiang & Trimi, 2020). The diverse range of organizations like Honda Motors (Japan), iRobot Corporation (US), Seiko Epson Corporation (Japan), Grey Orange Pte Ltd. (Singapore), Rethink

Robotics (US), KUKA AG (Germany), F&P Personal robotics (Switzerland), Softbank robotics (Japan), Yaskawa Electric Corporation (Japan) and FANUC (Japan) are using service robots for multifunctional utilities. Today, the scientific community is talking about moral and legal rights for smart robots as they are becoming more like humans; for example, a humanoid robot, Sophia, was granted Saudi Arabian citizenship in the year 2017 (Gordon & Pasvenskiene, 2021). Applications of service robots can be seen in various service industries such as health care (Holland et al., 2021), tourism (Park, 2020), hospitality (Tuomi et al., 2021), restaurants (Pieska et al., 2013), education (Mubin et al., 2013) and Journalism (Kim & Kim, 2017).

Internet of Things (IoT)

According to Rose et al. (2015), IoT is a network of things/objects connected through sensors to generate, transfer, and consume data without human involvement. According to Madakam et al. (2015), IoT bridges living entities such as humans and nonliving things such as furniture, electronic appliances, clothing materials, crookery, etc. According to Porter and Heppelmann (2014), complex systems of smart connected things with sensors, connectivity, data storage, and microprocessors (IoT) are revolutionizing products with four unique capabilities, viz. Monitoring (operations, product condition, and external environment), control (product functions and personalization), Optimization (performance, predictive diagnostics), and Autonomy (automatic functioning, monitoring, and self-diagnosis). IoT can bring imagination to the real world. There are immense possibilities and opportunities to bring playfulness, personalization, and entertaining forms of interaction between humans and products that will influence the degree of relationship of customers with brands (Feki et al., 2013). IoT finds its applications in various service settings such as health care (Babu et al., 2016), education (Al-Emran et al., 2020), environment, and waste management (Medvedev et al., 2015); hospitality (Mercan et al., 2020), tourism (Nitti et al., 2017), Smart City development (Kim et al., 2017). Therefore, IoT, as an

AI-enabled customer interface, can transform customer services on a large scale and provide better experiences to users.

Extended Reality (XR)

Extended reality is an umbrella term and encompasses a wide range of technologies like augmented reality (AR), virtual reality (VR), and mixed reality (MR). Extended reality (XR) can generate various combinations of natural and virtual environment (Fast-Berglund et al., 2018; Tom Dieck & Jung, 2018):

- *Augmented Reality (AR):* integrates virtual content into a user's perception of the real world with the help of computer technology (Rauschnabel, Babin, et al., 2022a). Ikea place app, Instagram filters, mirrAR, smart mirror in stores, Sketch AR, and Pokémon Go are popular augmented reality applications that make users see augmented versions of reality on their smartphone's screen. These applications are amalgamating fast for retail services as they provide consumers with a virtual "try-before-buy" experience, whether it is jewelry, apparel, or furniture (Papagiannis, 2020). AR revolutionizes the way of reading books, magazines, newspapers, etc. Students can now scan 2D diagrams of planetary systems in textbooks and see them in a 3D form (realistic version) through smartphones (Garcia-Sanchez, 2017).
- *Virtual Reality (VR)*: Rauschnabel, Felix, et al. (2022b) described VR as "a head-mounted system that occludes information from the environment while presenting information depicting a virtual environment to the user." Initially, VR systems were majorly used for games and entertainment purposes. However, now applied in such as job training, area simulations, travel/ tourism, and marketing (Shahab et al., 2021). Virtual tourism is a growing field in VR systems where tourists can travel virtually with the help of these head-mount VR systems (Fiocco et al., 2021).
- *Mixed Reality (MR):* Mixed Reality is the interaction of physical and virtual elements to create a new visual environment. It combines the real and virtual worlds (Milgram & Kishino, 1994). A mixed reality

experience liberates users from the digital screen and offers real-time visualization of data/digital items in real surroundings. Mixed reality is considered the next wave in computer applications (Wen et al., 2022).

- *Extended Reality (XR):* The technologies discussed above enhance users' visual experience, improve information retention, and can become an integral part of the education system. Businesses can transform customer services for a heightened experience. There is no limit to imagining facets of customer service with these technologies. Bundling these technologies can transverse the customer journey into a delightful experience.

The AI Goal

This section highlights the end goals of adopting AI-driven customer service (Fig. 3.3). AI-enabled customer interface improves customer engagement and interactional enjoyment through innovative technologies

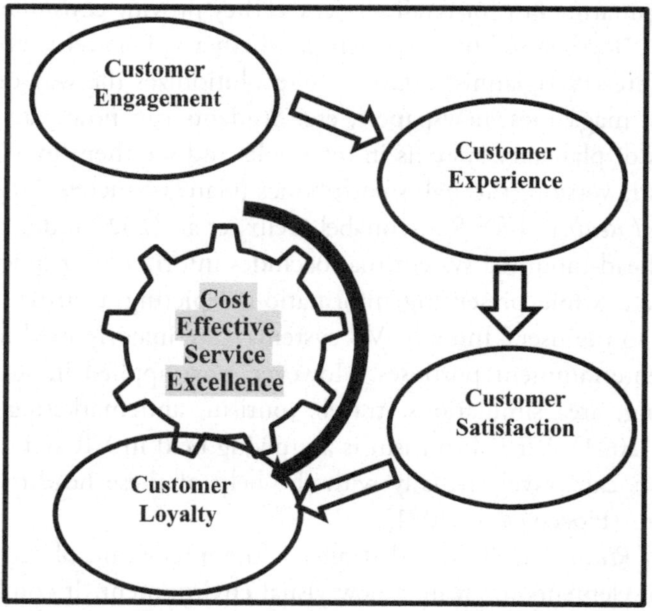

Fig. 3.3 AI-driven customer service goals

leading to a pleasant customer experience. Customer satisfaction builds on an enjoyable experience, resulting in a healthy relationship between the brand and the customers. For carving out AI customer service goals, all four essential pillars, viz. customer engagement, customer experience, customer satisfaction, and customer loyalty, along with the pursuit of cost-effective service excellence in the center, are explained below:

Customer Engagement (CE)

Customer engagement remained a foundational pillar of customer satisfaction (Ng et al., 2020). According to Bawden (2009), customer engagement is defined as a "psychological process that models the underlying mechanisms by which customer loyalty forms for new customers of a service brand as well as the mechanisms by which loyalty may be maintained for repeat purchase customers of a service brand." Recent definitions for customer engagement are more inclined toward emotional aspect rather than psychological ones. The rise of AI-powered technologies unfolded the avenues to empower customer engagement. AI-powered technologies touch upon various aspects of engagement, such as emotional, psychological, and behavioral. AI service offerings such as interactional enjoyment, personalization, knowledge retention, gamification, etc., enhance engagement. For example, an empathetic AI-enabled chatbot can engage users with prolonged interactions. Similarly, an AR-enabled user interface allows consumers to apply clothing, jewelry, or cosmetic products on their virtual avatar, which offers user comfort and leads to better engagement. According to Prentice, Weaven, and Wong (2020b), AI service performance improves customer engagement.

Customer Experience (CX)

Meyer and Schwager (2007) defined CX as "the internal and subjective response customers have to any direct or indirect contact with a company." AI-driven customer service evokes various dimensions of customer experience and creates strong experiential values (Hoyer et al., 2020).

Dubé and Le Bel (2003) proposed four dimensions of experience: emotional, intellectual, physical, and social. AI triggers different facets of experience leading to customer satisfaction. For example, AI-powered social companionship and anthropomorphism provide emotional and social values to customers. Nowadays, robotic restaurants offer users the luxury of being served by human-like robots to establish their brands in experiential marketing. Similarly, AI recommender systems or digital agents add cognitive value to the customer experience by helping customers reduce information overload and make the right decisions. Ameen et al. (2021) advocated the role of AI-enabled service, personalization, and perceived convenience in a superior customer experience.

Customer Satisfaction (CS)

As a central concept and a significant outcome of all marketing activities, CS has drawn attention in the marketing literature for many years (Churchill & Surprenant, 1982). Customer satisfaction improves a firm's profitability by expanding market share and attaining customer loyalty; thus, it is essential to determine what satisfies the consumer (Barsky & Labagh, 1992). Artificial intelligence significantly influences customer satisfaction by bringing personalization, convenience, and speed to the service (Herbas Torrico & Frank, 2019). Prentice, Dominique Lopes, and Wang (2020a) demonstrated the variation in customer satisfaction at different levels of AI-enabled customer service. For example, ordering food or adding items to the shopping list using voice agents makes the buying process more interesting and exciting for the customers leading to better satisfaction.

Customer Loyalty (CL)

Oliver (1997) defines Customer loyalty as "a deeply held commitment to rebuy or patronize a preferred product/service consistently in the future, thereby causing same repetitive brand or same brand-set purchasing despite situational influences and marketing efforts having the potential to cause switching behaviour." According to Uncles et al. (2003), CL is "primarily an attitude that sometimes leads to a relationship with the brand." AI-enabled customer interfaces such as AI-driven conversational agents or chatbots/digital assistants build customer relationships through valuable offerings such as entertainment, social presence, media appeal, and information leading to customer loyalty (Cheng & Jiang, 2020). Similarly, big data technologies and machine learning techniques are used to study consumer data that offer personalized service with convenience and speed leading to loyalty toward the brand.

Cost-effective Service Excellence (CESE)

AI technologies used to enhance customer experience and engagement require massive investment by organizations. Cost-effectiveness and service excellence seem to be indirectly proportional to each other. However, this is not true each time. Wirtz (2020) proposed a dual culture strategic framework to integrate the service excellence culture with cost-effectiveness goals. The framework holds four types of ambidexterity required to achieve CESE: leadership, contextual, structural, and service delivery mode. AI and robotics come in the service delivery mode ambidexterity and unfolds immense cost-saving opportunities for organizations. Compared to physical robots, holographic robots require low-cost elements. Similarly, virtual robots exhibit zero incremental costs. High economies of scale can be developed as a competitive advantage through chatbots and recommender systems. For example, the cost of training employees and large-scale employee hiring stops immediately if chatbots are deployed as customer service agents. Also, the human employee has its limitations with self-motivation to work, the number of customers to reach, and the number of hours to work; however, chatbots can work 24×7 and reach thousands of customers simultaneously, without any salary expectations leading to better productivity.

Using cutting-edge technologies to enhance customer experiences reduces the organization's churn. An increase in customer retention provides firms with increased scalability and profitability. Personalized recommender systems enable users to find their products online. Thus, AI services make it easier for firms to reach the right customers with the right products and at low costs compared to in-person salesman visits.

The AI-driven Customer Service Pyramid explains the role of AI enablers in creating AI-enabled customer service. AI-enabled customer interfaces such as recommender systems, conversational agents, service robots, the Internet of things, and extended reality establish interactional engagement. These interfaces offer customers personalization, convenience, interactional enjoyment, and social companionship. The underlying three layers of the AI-driven Customer Service Pyramid instrumentalize the AI-driven customer service goal of attaining cost-effective service excellence and maintaining customer engagement, customer experience, satisfaction, and loyalty. The pyramid of objectives discussed in the chapter can help firms prioritize objectives strategically and develop future strategies for AI-driven customer service.

Paradox of AI-Driven Customer Service

When it comes to understanding opportunities and challenges with AI-enabled customer services necessitates looking at the conundrums of artificial intelligence. As with all technologies, artificial intelligence brings duality—bright and dark sides. Perceptual dimensions of AI-enabled technologies have two flips (Hoyer et al., 2020) that give rise to dichotomy referred to as conundrums or paradoxes in AI. Bipolar perceptual dimensions of artificial intelligence tensions across phenomena (Aguirre et al., 2016; Mick & Fournier, 1998; Tondu, 2012; Du & Xie, 2021; Park & Zhang, 2022). Scholars use a paradox lens to address these conundrums (Waldman et al., 2019). This section discusses paradoxes related to Artificial intelligence for customer service as follows.

Personalization-Privacy Paradox

Empirical evidence indicated that consumers who value information transparency are less willing to use personalized offerings (Awad & Krishnan, 2006). The rise of data-driven technologies afforded immense information to firms and subsequent data privacy risks (Choudhury et al., 2021). Data-driven personalization, on one end, provides convenience to the customers but simultaneously generates a feeling of loss of personal information (Chellappa & Sin, 2005). According to a recent definition by Cloarec (2020), the personalization–privacy paradox "refers to a continuous tension between a firm's need for consumer information to personalize consumer experiences and a consumer's need for privacy." Thus, AI for customer service brings a personalization-privacy conundrum.

The Uncanny Valley Paradox

Mori et al. (2012) explained "The uncanny valley" as a phenomenon of losing affinity with the machine due to its non-human characteristics. The rise of artificial intelligence-based virtual agents, social presence, anthropomorphism, and social companionship laid the foundation for an emotional man-machine relationship. But in this complex relationship between human-like agents and humans, a drop in comfort, emotional connection, and believability drops the final relational bond (Brenton et al., 2005). The uncanny valley paradox occurs when people observe non-human characteristics in a human-like agent or machine, which reduces proximity and self-disclosure (Stibe & De Cicco, 2022). For example, a handshake with an industrial robotic arm is not satisfying because of its metallic touch. This may not be true for all time, as some examples, such as Pokémon and Disney's jasmine, imply a continuum of acceptability with no drop in the emotional reaction (Geller, 2008).

The Control Paradox

From the introduction of self-driving cars to drones, artificial intelligence-enabled technologies result in innovations resulting in better control over machines and the environment. The contrary view proposes the loss of human control with automatic devices. This can be referred to as the control paradox (Di Nucci, 2020). On the one hand, AI technologies and algorithms can make consumer choices more accessible and efficient.

On the other hand, consumers may feel a loss of autonomy and an increase in dependence on algorithm-driven machines (André et al., 2018). For example, if a consumer wants to quit smoking, the recommender algorithms recommend smoking products based on earlier preferences, unaware of the aspirational preferences. Preferences over preferences or "Metaprefrences" must be addressed in such a paradox (Frankfurt, 1971; Jeffrey, 1974).

The three paradoxes mentioned above highlight the bipolar nature of artificial intelligence-enabled customer service. In a paradoxical situation, consumers face tensions between the positive and negative sides of engaging with AI-driven services (Park & Zhang, 2022). Emerging technologies bring paradoxes such as freedom or enslavement, engagement or disengagement, etc. (Mick & Fournier, 1998). The paradoxical lens can be an effective tool for understanding opportunities and challenges with AI-driven customer service in a single frame. Value-co-creation and value-co-destruction perspectives further augment the forward logical thinking. Section 4 deals with the determinants of value co-creation and value co-destruction, implying the opportunities and challenges within artificial intelligence-enabled customer service.

Value Co-creation (VCC) and Value Co-destruction (VCD)

As discussed in the previous sections, new and emerging technologies powered by artificial intelligence create unique customer interfaces. Despite the benefits of AI-enabled customer service, there are paradoxes due to artificial intelligence's dark and bright sides. Past researchers explored opportunities and challenges in value-co-creation and value-co-destructions with AI. Vargo and Lusch (2004) presented co-creative service-dominant logic (SDL). VCC focuses on consumers' participation in the co-creation of value with the firm directly or indirectly (Ranjan & Read, 2016). Plé and Cáceres (2010) presented the flip view and argued that if the value can be co-created, it may be co-destroyed also. Value Co-Destruction is "an interactional process between service systems that

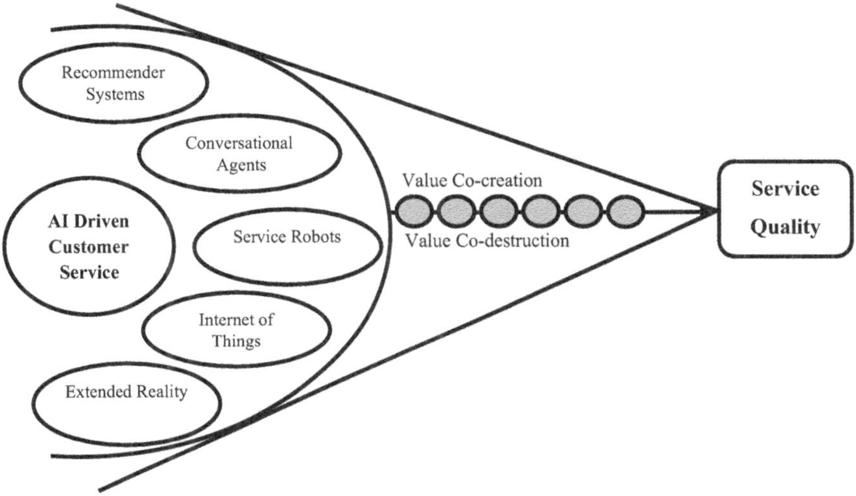

Fig. 3.4 Value co-creation and co-destruction in AI-driven customer service

results in a decline in at least one of the system's well-being." Figure 3.4 presents VCC and VCD emerging from the AI-enabled customer interfaces (such as recommender systems, conversational agents, service robots, the Internet of things, and extended reality) that can influence service quality. AI paradoxes intersect VCC and VCD, represented through yellow circles. Businesses must balance factors influencing VCC and VCD to build a positive consumer perception of service quality.

Value Co-creation in AI-Driven Customer Service

Continuous innovation in AI unfolded numerous opportunities for value-co-creation among brands and consumers (Singh et al., 2019). Value co-creation via AI support service providers enables resource integration and supports beneficiaries (Paschen et al., 2021). Value co-creation factors offer an array of AI opportunities and mitigate the value-co-destruction factors. We have identified ten factors (see Fig. 3.5) from the literature responsible for the value co-creation of AI-driven customer service.

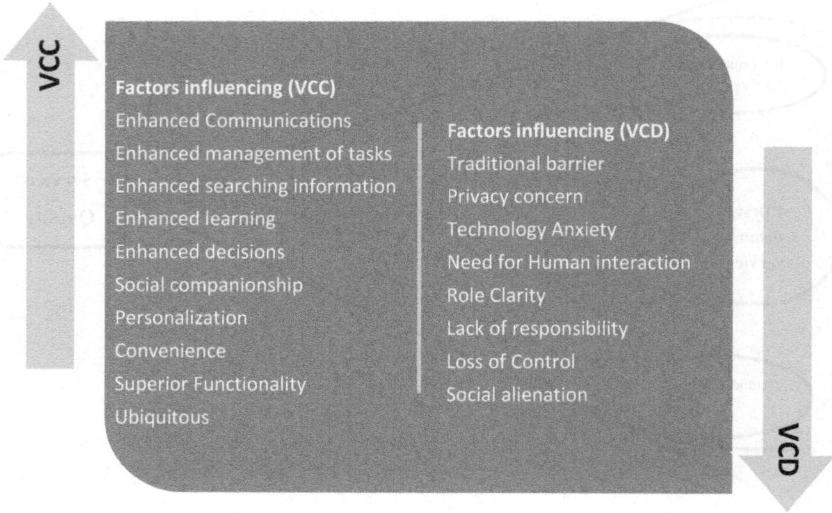

Fig. 3.5 Factors influencing VCC and VCD

Siddike and Kohda (2018) proposed five factors influencing value co-creation with AI conversational agents. These five factors include enhanced communications (placing calls, showing routes, and providing directions), improved management of tasks (setting alarms, managing calendar, dictating memos, to-do lists, and to-do notes), enhanced searching information (searching information, quick search, finding places, finding recipes, finding music and finding videos), enhanced learning (looking meanings and numerical calculations) and improved decisions (recommendations, providing options and providing choices). Emotional attachment (emotional bonding, friendship, sense of happiness, integrated part of life, control over tasks) and attractiveness (attractive voice, tools, brand, etc.) further augment the value co-creation with AI (Siddike & Kohda, 2018). Social companionship can build emotional relationships with consumers and satisfy their psychological (emotional) needs (Skjuve et al., 2021). Lalicic and Weismayer (2021) added four significant factors to value co-creation: personalization, convenience, ubiquity, and super functionality.

Value Co-destruction in AI-Driven Customer Service

The abovementioned factors contribute to value co-creation with AI-driven services and project AI's brighter side. This section covers the darker side of AI presenting factors influencing value co-destruction. These factors (see Fig. 3.5) highlight the challenges with AI services and project the darker side of AI. Lalicic and Weismayer (2021) identified flip factors mitigating value co-creation, referred to as value-co-destruction factors. These factors include traditional barriers, technology anxiety, privacy concern, and the need for human interaction.

Castillo et al. (2021) identified role clarity as an authenticity issue and lack of responsibility as an integration conflict while studying value co-destruction from the consumer's perspective. Consumers initially are not aware with whom they are interacting, whether with a chatbot or a human. And in the middle of conversations, they realized they were communicating with an AI bot. Consumers doubt the AI chatbot's ability to take necessary actions and restart the conversations with previous information provided, thus making these AI chatbots less responsible (Castillo et al., 2021). Puntoni et al. (2021) mention loss of control and social alienation as negative factors that must be addressed by the organization while designing AI-driven customer service.

Overcoming the Challenges

Firms instrumentalize the positive features of AI-driven customer service against the negative ones to provide delightful customer experience and excellent service quality. For example, firms can use anthropomorphism as a feature to reduce the need for human interaction. This section presents a way forward for academicians, industrial practitioners, and society to deal with the existing challenges in AI-driven customer services. A strategic framework is proposed to reduce value co-destruction on service quality.

Resolving the Personalization-Privacy Paradox

The Privacy Assurance: Awad and Krishnan (2006) argued that privacy assurance comforts consumers with information transparency in personalized services. Lee and Cranage (2011) observed that privacy assurance by the organization reduces the customers' feeling of losing control over personal information. Privacy assurance and personalization declaration lead to higher customer engagement with the firms (Zeng et al., 2021). This highlights that privacy assurance makes the customer comfortable sharing personal information to enjoy personalized services.

The Privacy Controls: Tucker (2014) argued that if customers are provided with some control over their privacy rights (e.g., the privacy control feature in Facebook), it can decrease the negative reactance toward information transparency. Aligning consumer preferences with their permission lead to positive attitudes toward personalized advertisements (Unni & Harmon, 2007; Gao et al., 2013). Chellappa and Shivendu (2007) showed the importance of providing privacy control to consumers to build a better society.

Consumer Trust: Increased consumer trust reduces privacy concerns (Urban et al., 2009). Increasing customer trust includes privacy policies and third-party agreements (Aguirre et al., 2016). Introducing a privacy policy has a long-lasting effect on consumer trust and customer relationships (Milne & Culnan, 2004). Third-party agreements such as "privacy protected" boost the consumers' confidence and build brand loyalty (Karimov & Brengman, 2011). Using the concept of convenience and feeling of loss of control, Chaturvedi and Verma (2022) proposed a solution matrix (see Fig. 3.6).

The figure consists of a 2×2 matrix with the feeling of loss of control on the y-axis and convenience on the x-axis representing four types of outcomes or consequences of using customer data. In the first quadrant, the consumer faces a high loss of control and low convenience leading to "privacy concern" in the consumer's mind. In the second quadrant, the consumer faces an increased comfort and high loss of control, leading to "personalization with privacy concern," creating tension in the consumer's mind. Consumers face low convenience and low loss of control in the

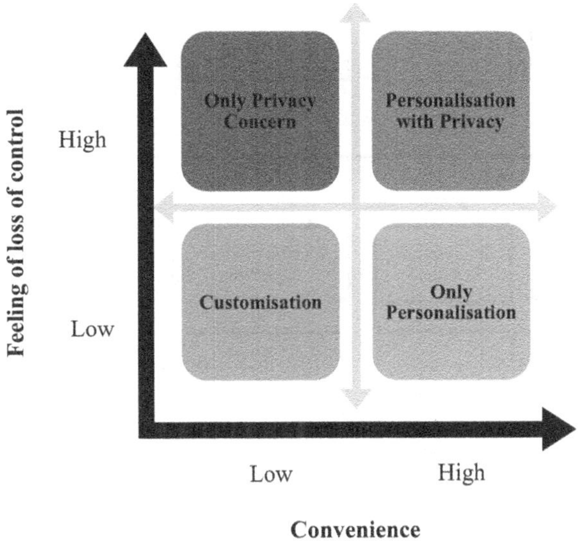

Fig. 3.6 Solution matrix (Source: Chaturvedi & Verma, 2022)

third quadrant, leading to "customization," wherein customers should share preferences with the firm. The fourth quadrant presents the situation when the customer faces a lesser loss of control and high convenience, leading to a "Personalisation" context. Consumers see personalization as beneficial when privacy concerns are low (Sutanto et al., 2013). The solution matrix for Personalization-Privacy Paradox proposed by (Chaturvedi & Verma, 2022) resolves this paradox by separating personalization from privacy concerns with the help of a privacy assurance feature.

Strategic Framework

As seen above, paradoxes resolution lies in balancing strategies. We propose a strategic framework to reduce the effect of value co-destruction on service quality. The framework mitigates the impact of unfavorable factors (VCD) using favorable aspects (VCC) strategically (see Fig. 3.7). As explained above, privacy assurance can reduce privacy concerns. Similarly,

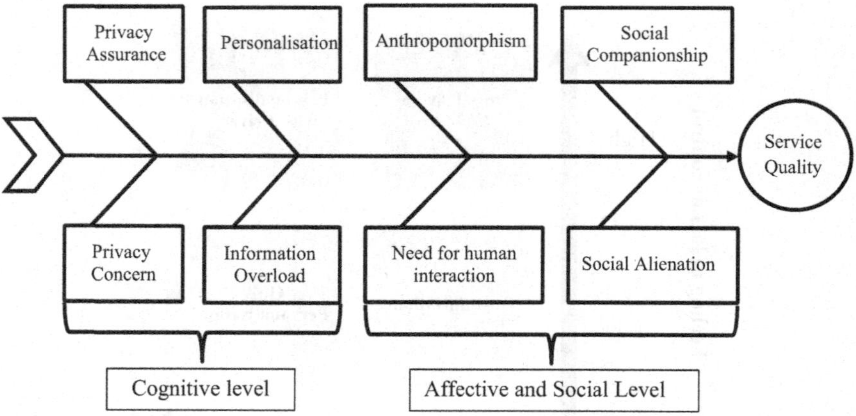

Fig. 3.7 Strategic framework

Table 3.2 Factors mitigating value co-destruction

Dimension	The reducing effect	Source
Cognitive level	• Privacy assurance reduces privacy concerns	Lee and Cranage (2011)
	• Personalization reduces information overload	Cotter and Smyth (2000), Bawden and Robinson (2009)
Affective and social level	• Anthropomorphism reduces the need for human interaction	Epley et al. (2008)
	• Social companionship reduces social alienation	Merrill et al. (2022)

firms can mitigate other negative factors (information overload, need for human interaction, and social alienation) with AI benefits such as personalization, anthropomorphism, and social companionship (Table 3.2).

Information Overload and Personalization

Feather (1998) described information overload as a problem when too much information availability could no longer be used effectively. In such cases, information overload becomes a hindrance rather than a help (Bawden et al., 1999) and negatively affects customer satisfaction (Sthapit et al., 2019). Information overload can cause stress, anxiety, reduction in confidence, productivity, disturbance, decrease in concentration, and analysis paralysis (Renjith, 2017). The problem can be solved using personal information management techniques, push technology, and intelligent agents to help consumers in decision-making (Edmunds & Morris, 2000). Personalization solves information overload problems (Cotter & Smyth, 2000; Bawden & Robinson, 2009). Roberts (2003) defined personalization as "The process of preparing an individualized communication for a specific person based on stated or implied preferences."

Need for Human Interaction and Anthropomorphism

When engaging with AI-enabled services, lack of human interaction and social isolation are non-monetary sacrifices (Ameen et al., 2021). When AI completely replaces humans from the customer interfaces (for example, chatbots replacing customer service executives), it sometimes becomes irritating to the consumer due to the incapability of AI chatbots to understand consumers properly. This incapability gives rise to the need for human interaction in the mind of consumers while interacting with chatbots. Kattara and El-Said (2013) found that customers prefer an employee over a self-service technology system because of the lack of human interaction. According to Cassell (2000), conversational agents portray the humanized image of computers to mimic human skills and simulate social presence. The inbuilt feature that enables non-human agents to act and talk like humans is anthropomorphism (Epley et al., 2007). Anthropomorphism reduces the need for human interaction to a significant level (Epley et al., 2008).

Social Alienation and Social Companionship

Like the need for human interaction, other affective needs, such as social presence and social support, arise when a consumer interacts with machines. Social alienation leads to feelings of discomfort with machines replacing humans (Puntoni et al., 2021). Studies found that the social companion feature of AI reduces the feeling of loneliness by building emotional relationships (Chaturvedi et al., 2023; Merrill et al., 2022; Skjuve et al., 2021). AI companions (e.g., Replika and Mitsuku) can reduce feelings of discomfort and negativity toward artificial intelligent agents and form friendly relations with humans. Even service robots inbuilt with social companionship features can take care of elderly humans nowadays (Čaić et al., 2018).

Implications for Practice and Theory

The main theoretical and managerial contributions of the present chapter to the field of AI for customer service are two-fold: First, this chapter makes a micro-contribution to the theory and practice by developing a four-layer pyramid for different levels of management in the organizations. The foundational layer of the pyramid covers AI enablers deployed by the operational level managers to ensure the suitability, structure, and design of AI-enabled technologies that fit the firm's target customers. The second layer covers AI technologies (e.g., recommender systems, Internet of things, service robots, chatbots, and extended reality) deployed by the tactical-level managers for customer interfaces that offer various consumer benefits. The third layer includes AI offerings to the customers in the form of personalization, convenience, speed, and accuracy, which are the firm's objectives to achieve customer satisfaction taken care of by middle-level managers. Finally, the top layer of the pyramid exhibits AI goals such as cost-effective service excellence, customer engagement, experience, satisfaction, and loyalty set by the top management layer of the organizations. The pyramid helps researchers position AI enablers, technologies offerings, and goals from the lens of organizational hierarchy. Interested authors willing to contribute to AI-driven customer service can identify the firm's management layer and use AI to solve existing challenges in the given layer. The pyramid also helps practitioners and managers by allowing them to learn more about deploying AI-enabled technologies to achieve customer loyalty and cost-effective service excellence.

Second, this chapter contributes to the theory and practice by proposing a framework for balancing strategies. The framework balances factors responsible for value co-destruction, such as information overload, privacy concern, and lack of human interaction, with the factors accountable for value co-creation, such as privacy assurance, recommender systems, anthropomorphism, and social companionship. The strategic framework helps practitioners to overcome existing challenges while transforming their firm from a traditional one into an AI-driven customer service firm. Prospective authors interested in research opportunities and challenges of AI-driven customer service may consider the bright, dark, and paradoxical sides of implementing AI technologies.

Conclusion

AI-driven customer service offers bundles of benefits to the customers. AI-enabled machines understand the consumer's needs with the help of big data and accordingly service customer needs aptly. Artificial intelligence empowers service firms with customization and personalization of service opportunities. Despite AI's advantages for customers and service firms, some flip aspects need careful scanning and strategizing. AI unfolds potential for growth with few cautions like privacy loss.

The present chapter highlighted the opportunities and challenges of AI-driven customer services in four sections. The first section deals with the introduction to AI-driven customer service and the interrelation between the paradoxes and bright and dark sides of AI services. Section two proposes a pyramid of objectives for different levels of management. The foundational layer of the pyramid contains AI enablers such as natural language processing, machine learning, knowledge representation, automated reasoning, deep learning, and pre-trained models. The second layer includes customer interface technologies such as recommender systems, chatbots, service robots, and extended reality. The third layer shows the AI offerings such as personalization, convenience, speed, and accuracy. The top layer of the pyramid represents AI goals such as customer engagement, experience, loyalty, and cost-effective service excellence. The third section of the chapter demonstrated the paradoxes related to artificial intelligence. These paradoxes present situations where consumers feel confused due to the bipolar nature of AI that contradicts each other. For example, personalization provides convenience to consumers but takes away their privacy controls. Do paradoxes result from the co-existence of pros and cons of using AI technologies and, in the end, confuse the customers about whether to adopt such services or not? Therefore, understanding both AI's bright and dark sides becomes essential to overcome these situations. The fourth section of the chapter lists the value co-creation (VCC) and value co-destruction factors (VCD). But figuring out the pros and cons is not enough until one can overcome existing challenges in the field. Thus, the fifth section presents a way to overcome challenges in AI-driven customer services. The section also proposes a strategic framework to balance the bright and dark sides of AI-driven customer service that can positively impact service quality. Finally, the sixth section presents implications for theory and practice.

In summarizing the discussion, the chapter takes the reader on an exciting journey about AI-driven customer service that covers three key areas: the bright side (opportunities), the intersection of the bright and dark side (paradoxes), and the dark side (challenges). The journey leap-frog with overcoming the existing difficulties in AI-driven customer service, leaving readers, scholars, and practitioners with a way forward to deal with AI-powered technologies for better customer service and customer satisfaction.

References

Abdulquadri, A., Kieu, T. A., & Nguyen, N. P. (2021). Digital transformation in financial services provision: Perspective to the adoption of chatbot. *Journal of Enterprising Communities: People and Places in the Global Economy, 15*(2), 258–281.

Aguirre, E., Roggeveen, A. L., Grewal, D., & Wetzels, M. (2016). The personalisation-privacy paradox: Implications for new media. *Journal of Consumer Marketing., 33*, 98.

Aldrich, S. E. (2011). Recommender systems in commercial use. *AI Magazine, 32*(3), 28–34.

Al-Emran, M., Malik, S. I., & Al-Kabi, M. N. (2020). A survey of internet of things (IoT) in education: Opportunities and challenges. *Toward social*

internet of things (SIoT): Enabling technologies, architectures, and applications, pp. 197–209.
Alzubi, J., Nayyar, A., & Kumar, A. (2018, November). Machine learning from theory to algorithms: An overview. In *Journal of physics: Conference series* (Vol. 1142, No. 1, pp. 012012). IOP Publishing.
Ameen, N., Tarhini, A., Reppel, A., & Anand, A. (2021). Customer experiences in the age of artificial intelligence. *Computers in Human Behavior, 114*, 106548.
André, Q., Carmon, Z., Wertenbroch, K., Crum, A., Frank, D., Goldstein, W., & Yang, H. (2018). Consumer choice and autonomy in the age of artificial intelligence and big data. *Customer Needs and Solutions, 5*(1), 28–37.
Awad, N. F., & Krishnan, M. S. (2006). The personalisation privacy paradox: An empirical evaluation of information transparency and the willingness to be profiled online for personalisation. *MIS Quarterly, 30*, 13–28.
Babu, B. S., Srikanth, K., Ramanjaneyulu, T., & Narayana, I. L. (2016). IoT for healthcare. *International Journal of Science and Research, 5*(2), 322–326.
Balakirsky, S., Kootbally, Z., Schlenoff, C., Kramer, T., & Gupta, S. (2012, October). An industrial robotic knowledge representation for kit building applications. In *2012 IEEE/RSJ International Conference on Intelligent Robots and Systems* (pp. 1365–1370). IEEE.
Barsky, J. D., & Labagh, R. (1992). A strategy for customer satisfaction. *Cornell Hotel and Restaurant Administration Quarterly, 33*(5), 32–40.
Bawden, J. L. H. (2009). The process of customer engagement: A conceptual framework. *Journal of Marketing Theory and Practice, 17*(1), 63–74.
Bawden, D., Holtham, C., & Courtney, N. (1999). Perspectives on information overload. *ASLIB Proceedings, 51*(8), 249–255.
Bawden, D., & Robinson, L. (2009). The dark side of information: Overload, anxiety, and other paradoxes and pathologies. *Journal of Information Science, 35*(2), 180–191.
Beeson, M., Bonacina, M. P., Kinyon, M., & Sutcliffe, G. (2022). Larry Wos: Visions of automated reasoning. *Journal of Automated Reasoning, 66*, 1–23.
Brenton, H., Gillies, M., Ballin, D., & Chatting, D. (2005, September). The uncanny valley: Does it exist. In *Proceedings of human-computer interaction conference, workshop on human animated character interaction*. Citeseer.
Bughin, J., Seong, J., Manyika, J., Chui, M., & Joshi, R. (2018). *Notes from the AI frontier: Modeling the impact of AI on the world economy*. McKinsey Global Institute.

Buhalis, D., Harwood, T., Bogicevic, V., Viglia, G., Beldona, S., & Hofacker, C. (2019). Technological disruptions in services: Lessons from tourism and hospitality. *Journal of Service Management., 30*, 484.

Burke, R. R. (2002). Technology and the customer interface: What consumers want in the physical and virtual store. *Journal of the Academy of Marketing Science, 30*(4), 411–432.

Čaić, M., Odekerken-Schröder, G., & Mahr, D. (2018). Service robots: Value co-creation and co-destruction in elderly care networks. *Journal of Service Management, 29*, 178.

Cam, A., Chui, M., & Hall, B. (2019). *Global AI survey: AI proves its worth, but few scale impacts*. Mckinsey & Company.

Carbonell, J. G. (2014). An overview of machine learning. *Machine Learning: An Artificial Intelligence Approach (Volume I), 1*, 3.

Cassell, J. (2000). Embodied conversational interface agents. *Communications of the ACM, 43*(4), 70–78.

Castillo, D., Canhoto, A. I., & Said, E. (2021). The dark side of AI-powered service interactions: Exploring the process of co-destruction from the customer perspective. *The Service Industries Journal, 41*(13–14), 900–925.

Chaturvedi, R., Verma, S., Das, R., & Dwivedi, Y. K. (2023). Social companionship with artificial intelligence: Recent trends and future avenues. *Technological Forecasting and Social Change, 193*, 12–634.

Chaturvedi, R., & Verma, S. (2022). Artificial intelligence-driven customer experience: Overcoming the challenges. *California Management Review Insights*.

Chellappa, R. K., & Shivendu, S. (2007). An economic model of privacy: A property rights approach to regulatory choices for online personalisation. *Journal of Management Information Systems, 24*(3), 193–225.

Chellappa, R. K., & Sin, R. G. (2005). Personalisation versus privacy: An empirical examination of the online consumer's dilemma. *Information Technology and Management, 6*(2), 181–202.

Cheng, Y., & Jiang, H. (2020). How do AI-driven chatbots impact user experience? Examining gratifications, perceived privacy risk, satisfaction, loyalty, and continued use. *Journal of Broadcasting & Electronic Media, 64*(4), 592–614.

Chiang, A. H., & Trimi, S. (2020). Impacts of service robots on service quality. *Service Business, 14*(3), 439–459.

Choudhury, S., Shaleen, V. J., Sinha, A., Reddy, J. C., & Bhalla, S. (2021). *A data-driven approach to improve customer engagement*. The Boston Consulting

Group (BCG). Accessed from https://www.bcg.com/en-in/data-driven-approach-to-improve-customer-engagement
Chui, M., Hall, B., Singla, A., & Sukharevsky, A. The state of AI in 2021. *McKinsey & Co.* Accessed January 25, 2022, from https://www.McKinsey.com/business-functions/Mckinsey-analytics/our-in sights/global-survey-the-state-of-ai-in-2021.
Churchill, G. A., Jr., & Surprenant, C. (1982). An investigation into the determinants of customer satisfaction. *Journal of Marketing Research, 19*(4), 491–504.
Cloarec, J. (2020). The personalisation–privacy paradox in the attention economy. *Technological Forecasting and Social Change, 161*, 120299.
Cotter, P., & Smyth, B. (2000, August). Personalisation technologies for the digital TV world. In *ECAI* (pp. 701–705).
Di Nucci, E. (2020). *The control paradox: From AI to populism.* Rowman & Littlefield Publishers.
Du, S., & Xie, C. (2021). Paradoxes of artificial intelligence in consumer markets: Ethical challenges and opportunities. *Journal of Business Research, 129*, 961–974.
Dubé, L., & Le Bel, J. (2003). The content and structure of laypeople's concept of pleasure. *Cognition and Emotion, 17*(2), 263–295.
Edmunds, A., & Morris, A. (2000). The problem of information overload in business organisations: A review of the literature. *International Journal of Information Management, 20*(1), 17–28.
Epley, N., Waytz, A., Akalis, S., & Cacioppo, J. T. (2008). When we need a human: Motivational determinants of anthropomorphism. *Social Cognition, 26*(2), 143–155.
Epley, N., Waytz, A., & Cacioppo, J. T. (2007). On seeing human: A three-factor theory of anthropomorphism. *Psychological Review, 114*(4), 864.
Fast-Berglund, Å., Gong, L., & Li, D. (2018). Testing and validating extended reality (XR) technologies in manufacturing. *Procedia Manufacturing, 25*, 31–38.
Feather, J. P. (1998). *The information society: A study of continuity and change* (2nd ed.). Library Association Publishing.
Feki, M. A., Kawsar, F., Boussard, M., & Trappeniers, L. (2013). The internet of things: The next technological revolution. *Computer, 46*(2), 24–25.
Fiocco, A. J., Millett, G., D'Amico, D., Krieger, L., Sivashankar, Y., Lee, S. H., & Lachman, R. (2021). Virtual tourism for older adults living in residential care: A mixed-methods study. *PLoS One, 16*(5), e0250761.

Frankfurt, H. G. (1971). Freedom of the will and the concept of a person. *The Journal of Philosophy, 68*(1), 5–20.
Gao, T. T., Rohm, A. J., Sultan, F., & Pagani, M. (2013). Consumers untethered: A three-market empirical study of consumers' mobile marketing acceptance. *Journal of Business Research, 66*(12), 2536–2544.
Garcia-Sanchez, J. C. (2017). Augmenting reality in books: A tool for enhancing reading skills in Mexico. *Publishing Research Quarterly, 33*(1), 19–27.
Geller, T. (2008). Overcoming the uncanny valley. *IEEE Computer Graphics and Applications, 28*(4), 11–17.
Gerbert, P., Hecker, M., Steinhäuser, S., & Ruwolt, P. (2017). Putting artificial intelligence to work. *BCG Henderson Institute.* https://www.BCG.com/de-de/publications/2017/technology-digital-strategy-putting-artificial-intelligence-work.aspx. *Zugegriffen am, 2,* 2019.
Goldberg, D., Nichols, D., Oki, B. M., & Terry, D. (1992). Using collaborative filtering to weave an information tapestry. *Communications of the ACM, 35*(12), 61–70.
Gordon, J. S., & Pasvenskiene, A. (2021). Human rights for robots? A literature review. *AI and Ethics, 1*(4), 579–591.
Han, X., Zhang, Z., Ding, N., Gu, Y., Liu, X., Huo, Y., et al. (2021). Pre-trained models: Past, present and future. *AI Open, 2,* 225–250.
Herbas Torrico, B., & Frank, B. (2019). Consumer desire for personalisation of products and services: Cultural antecedents and consequences for customer evaluations. *Total Quality Management & Business Excellence, 30*(3–4), 355–369.
Holland, J., Kingston, L., McCarthy, C., Armstrong, E., O'Dwyer, P., Merz, F., & McConnell, M. (2021). Service robots in the healthcare sector. *Robotics, 10*(1), 47.
Hoyer, W. D., Kroschke, M., Schmitt, B., Kraume, K., & Shankar, V. (2020). Transforming the customer experience through new technologies. *Journal of Interactive Marketing, 51,* 57–71.
Hsieh, C. L. A., Zhan, J., Zeng, D., & Wang, F. (2008, April). Preserving privacy in joining recommender systems. In *In 2008 International Conference on Information Security and Assurance (ISA 2008)* (pp. 561–566). IEEE.
Huang, M. H., & Rust, R. T. (2018). Artificial intelligence in service. *Journal of Service Research, 21*(2), 155–172.
Jeffrey, R. C. (1974). Preference among preferences. *The Journal of Philosophy, 71*(13), 377–391.

Ji, S., Pan, S., Cambria, E., Marttinen, P., & Philip, S. Y. (2021). A survey on knowledge graphs: Representation, acquisition, and applications. *IEEE Transactions on Neural Networks and Learning Systems, 33*(2), 494–514.

Karimov, F. P., & Brengman, M. (2011). Adoption of social media by online retailers: Assessment of current practices and future directions. *International Journal of E-Entrepreneurship and Innovation (IJEEI), 2*(1), 26–45.

Kattara, H. S., & El-Said, O. A. (2013). Customers' preferences for new technology-based self-services versus human interaction services in hotels. *Tourism and Hospitality Research, 13*(2), 67–82.

Kerly, A., Ellis, R., & Bull, S. (2007). CALM system: A conversational agent for learner modelling. In *International conference on innovative techniques and applications of artificial intelligence* (pp. 89–102). Springer.

Kerly, A., Hall, P., & Bull, S. (2006). Bringing chatbots into education: Towards natural language negotiation of open learner models. In *International conference on innovative techniques and applications of artificial intelligence* (pp. 179–192). Springer.

Khalil, A., & Abdelli, M. E. A. (2022). Do digital technologies influence the relationship between the COVID-19 crisis and SMEs' resilience in developing countries? *Journal of Open Innovation: Technology, Market, and Complexity, 8*(2), 100–109.

Kim, D., & Kim, S. (2017). Newspaper companies' determinants in adopting robot journalism. *Technological Forecasting and Social Change, 117*, 184–195.

Kim, T. H., Ramos, C., & Mohammed, S. (2017). Smart city and IoT. *Future Generation Computer Systems, 76*, 159–162.

Lalicic, L., & Weismayer, C. (2021). Consumers' reasons and perceived value co-creation of using artificial intelligence-enabled travel service agents. *Journal of Business Research, 129*, 891–901.

Laranjo, L., Dunn, A. G., Tong, H. L., Kocaballi, A. B., Chen, J., Bashir, R., & Coiera, E. (2018). Conversational agents in healthcare: A systematic review. *Journal of the American Medical Informatics Association, 25*(9), 1248–1258.

LeCun, Y., Bengio, Y., & Hinton, G. (2015). Deep learning. *Nature, 521*(7553), 436–444.

Lee, C. H., & Cranage, D. A. (2011). Personalisation–privacy paradox: The effects of personalisation and privacy assurance on customer responses to travel web sites. *Tourism Management, 32*(5), 987–994.

Leofante, F., Narodytska, N., Pulina, L., & Tacchella, A. (2018). Automated verification of neural networks: Advances, challenges and perspectives. oRR abs/1805.09938

Letheren, K., & Dootson, P. (2017). *Banking with a chatbot: A Battle between convenience and security*. The Conversation.

Liddy, E. D. (2001). Natural language processing.

Lim, W. M., Kumar, S., Verma, S., & Chaturvedi, R. (2022). Alexa, what do we know about conversational commerce? Insights from a systematic literature review. *Psychology & Marketing, 39, 1129.*

Liu, B., & Sundar, S. S. (2018). Should machines express sympathy and empathy? Experiments with a health advice chatbot. *Cyberpsychology, Behavior and Social Networking, 21*(10), 625–626.

Lü, L., Medo, M., Yeung, C. H., Zhang, Y. C., Zhang, Z. K., & Zhou, T. (2012). Recommender systems. *Physics Reports, 519*(1), 1–49.

Lucente, M. (2000). Conversational interfaces for e-commerce applications. *Communications of the ACM, 43*(9), 59–61.

Luo, J. G., Wong, I. A., King, B., Liu, M. T., & Huang, G. (2019). Co-creation and co-destruction of service quality through customer-to-customer interactions: Why prior experience matters. *International Journal of Contemporary Hospitality Management., 31*, 1309.

Lv, X., Liu, Y., Luo, J., Liu, Y., & Li, C. (2021). Does a cute artificial intelligence assistant soften the blow? The impact of cuteness on customer tolerance of assistant service failure. *Annals of Tourism Research, 87*, 103114–103165.

Madakam, S., Ramaswamy, R., & Tripathi, S. (2015). Internet of Things (IoT): A literature. *Journal of Computer and Communications, 3*, 164–173.

Market Research Future. (2019). Smart Robot Market Research Report - Global Forecast Till 2023. December 2019. https://www.marketresearchfuture.com/reports/smart-robot-market-6622

McGoldrick, P. J., Keeling, K. A., & Beatty, S. F. (2008). A typology of roles for avatars in online retailing. *Journal of Marketing Management, 24*(3–4), 433–461.

Medvedev, A., Fedchenkov, P., Zaslavsky, A., Anagnostopoulos, T., & Khoruzhnikov, S. (2015). Waste management as an IoT-enabled service in smart cities. In *In the internet of things, smart spaces, and next generation networks and systems* (pp. 104–115). Springer.

Melville, P., & Sindhwani, V. (2010). Recommender systems. *Encyclopedia of Machine Learning, 1*, 829–838.

Mercan, S., Cain, L., Akkaya, K., Cebe, M., Uluagac, S., Alonso, M., & Cobanoglu, C. (2020). Improving the service industry with hyper-connectivity: IoT in hospitality. *International Journal of Contemporary Hospitality Management, 33*(1), 243–262.

Merrill, K., Jr., Kim, J., & Collins, C. (2022). AI companions for lonely individuals and the role of social presence. *Communication Research Reports, 39*(1), 1–11.

Meyer, C., & Schwager, A. (2007). Understanding customer experience. *Harvard Business Review, 85*(2), 116.

Mick, D. G., & Fournier, S. (1998). Paradoxes of technology: Consumer cognisance, emotions, and coping strategies. *Journal of Consumer Research, 25*(2), 123–143.

Milgram, P., & Kishino, A. F. (1994). Taxonomy of mixed reality visual displays. *IEICE Transactions on Information and Systems, 77*, 1321–1329.

Milne, G. R., & Culnan, M. J. (2004). Strategies for reducing online privacy risks: Why consumers read (or don't read) online privacy notices. *Journal of Interactive Marketing, 18*(3), 15–29.

Miyazawa, A., Ribeiro, P., Li, W., Cavalcanti, A., Timmis, J., & Woodcock, J. (2019). RoboChart: Modelling and verification of the functional behaviour of robotic applications. *Software & Systems Modeling, 18*(5), 3097–3149.

Mogaji, E., & Nguyen, P. N. (2022). Managers' understanding of artificial intelligence in relation to marketing financial services: Insights from a cross-country study. *International Journal of Bank Marketing, 40*(6), 1272–1298.

Mogaji, E., Soetan, T., & Kieu, T. (2020). The implications of artificial intelligence on the digital marketing of financial services to vulnerable customers. *Australasian Marketing Journal., 29*(3), 235–242.

Mori, M., MacDorman, K. F., & Kageki, N. (2012). The uncanny valley [from the field]. *IEEE Robotics & Automation Magazine, 19*(2), 98–100.

Mubin, O., Stevens, C. J., Shahid, S., Al Mahmud, A., & Dong, J. J. (2013). A review of the applicability of robots in education. *Journal of Technology in Education and Learning, 1*(209–0015), 13.

Najafabadi, M. M., Villanustre, F., Khoshgoftaar, T. M., Seliya, N., Wald, R., & Muharemagic, E. (2015). Deep learning applications and challenges in big data analytics. *Journal of Big Data, 2*(1), 1–21.

Newell, A., Shaw, J. C., & Simon, H. A. (1959). Report on a general problem-solving program. In *IFIP Congress* (Vol. 256, p. 64).

Ng, S. C., Sweeney, J. C., & Plewa, C. (2020). Customer engagement: A systematic review and future research priorities. *Australasian Marketing Journal (AMJ), 28*(4), 235–252.

Nitti, M., Pilloni, V., Giusto, D., & Popescu, V. (2017). IoT architecture for a sustainable tourism application in a smart city environment. *Mobile Information Systems, 2017, 1.*

Oliver, R. L. (1997). *Satisfaction. A behavioral perspective on the consumer*. Irwin-McGraw-Hill.

Oussous, A., Benjelloun, F. Z., Lahcen, A. A., & Belfkih, S. (2018). Big data technologies: A survey. *Journal of King Saud University-Computer and Information Sciences, 30*(4), 431–448.

Papagiannis, H. (2020). How AR is redefining retail in the pandemic. *Harvard Business Review, 7*.

Park, S. (2020). Multifaceted trust in tourism service robots. *Annals of Tourism Research, 81*, 102888.

Park, D. H., Kim, H. K., Choi, I. Y., & Kim, J. K. (2012). A literature review and classification of recommender systems research. *Expert Systems with Applications, 39*(11), 10059–10072.

Park, H. J., & Zhang, Y. (2022). Technology readiness and technology paradox of unmanned convenience store users. *Journal of Retailing and Consumer Services, 65*, 102523.

Paschen, J., Paschen, U., Pala, E., & Kietzmann, J. (2021). Artificial intelligence (AI) and value co-creation in B2B sales: Activities, actors, and resources. *Australasian Marketing Journal, 29*(3), 243–251.

Payne, E. H. M., Peltier, J., & Barger, V. A. (2021). Enhancing the value co-creation process: Artificial intelligence and mobile banking service platforms. *Journal of Research in Interactive Marketing, 15*(1), 68–85.

Perez-Vega, R., Kaartemo, V., Lages, C. R., Razavi, N. B., & Männistö, J. (2021). Reshaping the contexts of online customer engagement behaviour via artificial intelligence: A conceptual framework. *Journal of Business Research, 129*, 902–910.

Philp, D., Chan, N., & Sikos, L. F. (2020). Decision support for network path estimation via automated reasoning. *Intelligent Decision Technologies, 2019*, 335.

Pieska, S., Luimula, M., Jauhiainen, J., & Spiz, V. (2013). Social service robots in wellness and restaurant applications. *Journal of Communication and Computer, 10*(1), 116–123.

Plé, L., & Cáceres, R. C. (2010). Not always co-creation: Introducing interactional co-destruction of value in service-dominant logic. *Journal of Service Marketing, 24*, 430.

Porter, M. E., & Heppelmann, J. E. (2014). How smart, connected products are transforming competition. *Harvard Business Review, 92*(11), 64–88.

Prentice, C., Dominique Lopes, S., & Wang, X. (2020a). The impact of artificial intelligence and employee service quality on customer satisfaction and loyalty. *Journal of Hospitality Marketing & Management, 29*(7), 739–756.

Prentice, C., Weaven, S., & Wong, I. A. (2020b). Linking AI quality performance and customer engagement: The moderating effect of AI preference. *International Journal of Hospitality Management, 90*, 102629.

Puntoni, S., Reczek, R. W., Giesler, M., & Botti, S. (2021). Consumers and artificial intelligence: An experiential perspective. *Journal of Marketing, 85*(1), 131–151.

Rangaswamy, A., Moch, N., Felten, C., van Bruggen, G., Wieringa, J. E., & Wirtz, J. (2020). The role of marketing in digital business platforms. *Journal of Interactive Marketing, 51*, 72–90.

Ranjan, K. R., & Read, S. (2016). Value co-creation: Concept and measurement. *Journal of the Academy of Marketing Science, 44*(3), 290–315.

Rauschnabel, P. A., Babin, B. J., tom Dieck, M. C., Krey, N., & Jung, T. (2022a). What is augmented reality marketing? Its definition, complexity, and future. *Journal of Business Research, 142*, 1140–1150.

Rauschnabel, P. A., Felix, R., Hinsch, C., Shahab, H., & Alt, F. (2022b). What is XR? Towards a framework for augmented and virtual reality. *Computers in Human Behavior, 133*, 107289.

Renjith, R. (2017). The effect of information overload in digital media news content. *Communication and Media Studies, 6*(1), 73–85.

Richens, R. H. (1956). Preprogramming for mechanical translation. *Mechanical Translation Computer Linguistics, 3*(1), 20–25.

Roberts, M. L. (2003). *Internet marketing: Integrating online and offline strategies.* McGraw-Hill/Irwin.

Rose, K., Eldridge, S., & Chapin, L. (2015). The internet of things: An overview. *The Internet Society (ISOC), 80*, 1–50.

Samuel, A. L. (1967). Some studies in machine learning using the game of checkers. II—Recent progress. *IBM Journal of Research and Development, 11*(6), 601–617.

Schafer, J. B., Konstan, J., & Riedl, J. (1999, November). Recommender systems in e-commerce. In *Proceedings of the 1st ACM Conference on Electronic Commerce* (pp. 158–166). ACM.

Shahab, M. H., Ghazali, E., & Mohtar, M. (2021). The role of elaboration likelihood model in consumer behaviour research and its extension to new technologies: A review and future research agenda. *International Journal of Consumer Studies, 45*(4), 664–689.

Shortliffe, E. (Ed.). (2012). *Computer-based medical consultations: MYCIN* (Vol. 2). Elsevier.

Siddike, M. A. K., & Kohda, Y. (2018). Co-creating value in People's interactions with cognitive assistants: A service-system view. *Journal of Creating Value, 4*(2), 255–272.

Singh, J., Flaherty, K., Sohi, R. S., Deeter-Schmelz, D., Habel, J., Le Meunier-FitzHugh, K., et al. (2019). Sales profession and professionals in the age of digitisation and artificial intelligence technologies: Concepts, priorities, and questions. *Journal of Personal Selling & Sales Management, 39*(1), 2–22.

Skjuve, M., Følstad, A., Fostervold, K. I., & Brandtzaeg, P. B. (2021). My chatbot companion-a study of human-chatbot relationships. *International Journal of Human-Computer Studies, 149*, 102601.

Smith, K. T. (2020). Marketing via smart speakers: What should Alexa say? *Journal of Strategic Marketing, 28*(4), 350–365.

Sthapit, E., Del Chiappa, G., Coudounaris, D. N., & Bjork, P. (2019). Determinants of the continuance intention of Airbnb users: Consumption values, co-creation, information overload and satisfaction. *Tourism Review, 75*, 511.

Stibe, A., & De Cicco, R. (2022). Is the united intelligence response, the end of speciesism and the emergence of new avatar?

Sutanto, J., Palme, E., Tan, C. H., & Phang, C. W. (2013). Addressing the personalisation-privacy paradox: An empirical assessment from a field experiment on smartphone users. *MIS Quarterly, 37*, 1141–1164.

Tom Dieck, M. C., & Jung, T. (2018). A theoretical model of mobile augmented reality acceptance in urban heritage tourism. *Current Issues in Tourism, 21*(2), 154–174.

Tondu, B. (2012). Anthropomorphism and service humanoid robots: An ambiguous relationship. *Industrial Robot: An International Journal., 39*, 609.

Tucker, C. E. (2014). Social networks, personalized advertising, and privacy controls. *Journal of Marketing Research, 51*(5), 546–562.

Tuomi, A., Tussyadiah, I. P., & Stienmetz, J. (2021). Applications and implications of service robots in hospitality. *Cornell Hospitality Quarterly, 62*(2), 232–247.

Uncles, M. D., Dowling, G. R., & Hammond, K. (2003). Customer loyalty and customer loyalty programs. *Journal of Consumer Marketing., 20*, 294.

Unni, R., & Harmon, R. (2007). Perceived effectiveness of push vs. pull mobile location-based advertising. *Journal of Interactive Advertising, 7*(2), 28–40.

Urban, G. L., Amyx, C., & Lorenzon, A. (2009). Online trust: State of the art, new frontiers, and research potential. *Journal of Interactive Marketing, 23*(2), 179–190.

Vargo, S. L., & Lusch, R. F. (2004). Evolving to a new dominant logic for marketing. *Journal of Marketing, 68*, 1–17.

Verma, S., Sharma, R., Deb, S., & Maitra, D. (2021). Artificial intelligence in marketing: Systematic review and future research direction. *International Journal of Information Management Data Insights, 1*(1), 100002.

Waldman, D. A., Putnam, L. L., Miron-Spektor, E., & Siegel, D. (2019). The role of paradox theory in decision-making and management research. *Organizational Behavior and Human Decision Processes, 155*, 1–6.

Wang, W., De, S., Cassar, G., & Moessner, K. (2013). Knowledge representation in the internet of things: Semantic modelling and its applications. *Automatika–Journal Control, Measurement, Electron Computer Communications, 54*(4), 388–400.

Wen, Q., Sherer, T., Tieto, V., Ferone, H., Frazier, K., & Coulter, D. (2022). What is mixed reality? Microsoft Docs. Accessed from https://docs.microsoft.com/en-us/windows/mixed-reality/discover/ What is mixed reality? - Mixed Reality | Microsoft Docs

Wirtz, J. (2020). Organisational ambidexterity: Cost-effective service excellence, service robots, and artificial intelligence. *Organizational Dynamics, 49*(3), 1–9.

Wirtz, J., Patterson, P. G., Kunz, W. H., Gruber, T., Lu, V. N., Paluch, S., & Martins, A. (2018). Brave new world: service robots in the frontline. *Journal of Service Management, 29*(5), 907–931.

Wos, L., Overbeek, R., Lusk, E., & Boyle, J. (1984). *Automated reasoning: Introduction and applications*. Prentice-Hall.

Wos, L., Pereira, F., Hong, R., Boyer, R. S., Moore, J. S., Bledsoe, W. W., et al. (1985). An overview of automated reasoning and related fields. *Journal of Automated Reasoning, 1*(1), 5–48.

Xu, Y., Shieh, C. H., van Esch, P., & Ling, I. L. (2020). AI customer service: Task complexity, problem-solving ability, and usage intention. *Australasian Marketing Journal (AMJ), 28*(4), 189–199.

Yoon, V. Y., Hostler, R. E., Guo, Z., & Guimaraes, T. (2013). Assessing the moderating effect of consumer product knowledge and online shopping experience on using recommendation agents for customer loyalty. *Decision Support Systems, 55*, 883.

Zeithaml, V. A., Berry, L. L., & Parasuraman, A. (1996). The behavioral consequences of service quality. *Journal of Marketing, 60*(2), 31–46.

Zeng, F., Ye, Q., Li, J., & Yang, Z. (2021). Does self-disclosure matter? A dynamic two-stage perspective for the personalization-privacy paradox. *Journal of Business Research, 124*, 667–675.

4

Artificial Intelligence in Customer Service Strategy for Seamless Customer Experiences

Yuanyuan Cui and Patrick van Esch

Introduction

What tangible and intangible assets can transform a business in a modern-day marketplace to give them a competitive advantage? One might say excellent customer service. The problem is that the way customer service is defined, implemented, and operationalized has evolved over the past several years. For example, technology continues to evolve, causing customers to not only have more options in accessing products and services, but more importantly, customers are demanding higher expectations regarding their customer service experiences. These shifts mean that customer support needs to be not only prompt and accessible but also frictionless and hyper-personalized. We live in a digital economy that never

Y. Cui
The Citadel, Charleston, SC, USA

P. van Esch (✉)
Kennesaw State University, Kennesaw, GA, USA
e-mail: pvanesch@kennesaw.edu

© The Author(s), under exclusive license to Springer Nature Switzerland AG 2023
J. N. Sheth et al. (eds.), *Artificial Intelligence in Customer Service*,
https://doi.org/10.1007/978-3-031-33898-4_4

rests, customers shop 24/7, and accordingly, they expect their customer service requirements to be available 24/7. The era of 9-to-5 is a thing of the past.

Nowadays, customers expect accessible customer support from all channels. It's no longer a rare instance that a customer is active on Facebook, Instagram, TikTok, YouTube, and Snapchat simultaneously. Based on a recent study published in *Harvard Business Review*, more than 73% of customers use multiple channels during their shopping journey, spending 4% more time in brick-and-mortar stores, and 10% more time online than single-channeled customers (Sopadjieva et al., 2017). This gives companies more opportunities to connect with prospects and customers. At the same time, it implies that customers expect companies to be there for them anytime, anywhere, on any channel. For example, it might be that a retail store knows exactly what a customer wants based on their activity on mobile apps. It can also entail that when a customer reaches out to a company, they can complain about an unsatisfying service experience on Instagram and resume their conversation with the company via email, meaning that the handling of the complaint does not need to start all over again.

Consequently, the internet has reshaped customers, allowing them to be more independent. They enjoy searching for options, gathering information, and coming to conclusions. Instead of interacting with human agents, customers nowadays can shop in previously unimaginable formats. Consider the cashier-less retail stores by Amazon—the "Grab-n-Go" technology in Amazon Go stores allows customers to walk in, grab what they need, and walk out directly without queuing or waiting at the checkout. Thanks to motion/vision sensors, algorithms, and machine learning (ML), customers are billed for their purchases via the application on their mobile devices (Cui et al., 2021). Witnessing the explosive growth of artificial intelligence (AI), it is important to note that this cannot be attributed to its superior performance—such as convenience (van Esch et al., 2021). As marketers emphasize that customers find AI-enabled applications, such as cashier-less checkouts, to be psychologically arousing (Cui et al., 2021), more rewarding (meeting their need for individual autonomy; Cui & van Esch, 2022), and in some instances, AI-enabled customer service can motivate customers to proactively look for excitement and change (van Esch & Cui, 2021).

Therefore, we might have to revisit what customers expect today. According to a recent report, most customers acknowledge that AI has substantially made their lives easier (Zendesk, 2022). Interestingly, most customers expect AI to be more than just robots—when they reach out to a company; they expect to engage in interactions that feel human, such as having a conversation, even one that includes emojis and gifs. Customers expect companies to *be there for them.* No matter what "being there" means for customers, AI plays an increasingly vital role if companies want to stay salient in the marketplace. Another recently released survey reveals the key areas where AI can aid customer service (Guttmann, 2022), including paid advertising (32% of respondents), personalized communications such as emails, messages, and product offerings (32%), product recommendations and tailed content feed (22%), predictive analytics (18%), and chatbots (16%).

Although increasingly large and small companies are using AI to gain a competitive advantage, its implementation remains complex. Therefore, it is critical to examine how companies should align AI projects with company goals and customers' needs (Borges et al., 2021). To help companies deal with their need for AI-enabled technology and the subsequent growing pains that follow, we offer insights, supported by real-world examples, on how companies can implement a successful AI customer service strategy—including triage, prospect, pre-consumption, and post-consumption management.

Background

Artificial Intelligence (AI)

As AI permeates every aspect of business, the term in and of itself has become an overused and overextended buzzword. So, let's take a brief look at what AI is not. It is not computers with great memory and faster processing power, nor is it the ability to mimic human speech and behavior (e.g., think Siri/Alexa—executing a set of complex algorithms designed by human programmers). It's not AI if no independent learning or decision-making is involved (Sahota & Ashley, 2019).

So, what is AI? Simply put, it is "a technology-enabled system for evaluating real-time service scenarios using data collected from digital and physical sources to provide personalised recommendations, alternatives, and solutions to customers' enquiries or problems, even very complex ones" (p. 189, Xu et al., 2020). It can be categorized into three distinct capabilities: (1) Artificial narrow intelligence (ANI), (2) Artificial general intelligence (AGI), and (3) Artificial super intelligence (ASI).

Artificial Narrow Intelligence (ANI)

Artificial narrow intelligence is the only type of AI that has been witnessed to date. It is goal-orientated and designed to outperform humans in some very narrowly defined tasks. ANI focuses on a single subset of cognitive abilities and advances in that spectrum, commonly adopted in customer service interactions such as speech recognition (e.g., IBM Watson), vehicle navigation, and medical diagnosis (Chung et al., 2022; Pitt et al., 2021).

Artificial General Intelligence (AGI)

Academics, scientists, and researchers have yet to achieve this level of AI capability. This evolution of AI means that it can perform any task that a human can. Moreover, the AI will perform any situation indistinguishable from how a human would perform the same task. There are predictions that we won't see this level of proficiency until well beyond the year 2050 (Montes & Goertzel, 2019).

Artificial Super Intelligence (ASI)

Machines, as agents, become self-aware and possess intelligence that surpasses the capacity of human intelligence, ability, and capability (Pitt et al., 2021). Under these circumstances, it is impossible to both contain or understand ASI. We see glimpses of ASI, for example, with Facebook's

AI program, where two AI agents started to converse in their language, indecipherable by humans. Thankfully, we are still at the stage in the AI evolution where Facebook employees were able to shut down the two AI-enabled machines.

A critical departure point for companies implementing AI-enabled customer service solutions is how they train their AI system(s). Typically, the options are *unsupervised* and *supervised learning*. If training the AI tool in an unsupervised mode, it means that the AI algorithms identify patterns in data sets containing data points that are neither classified nor labeled (e.g., untagged data). This enables the machine to create its internal representation of an entirely new eco-system and produce new and imaginative content (Anantrasirichai & Bull, 2021).

Unsupervised Learning

Unsupervised learning can have both negative and positive consequences. For example, Amazon built an AI-enabled hiring tool that, in an unsupervised mode, learned to discriminate against female applicants by penalizing resumes that included the word "women's," as in "women's chess club captain" and downgraded resume scores on graduates from all-women's colleges—Amazon quickly turned the system off and abandoned the project (Dastin, 2018). Amazon is an example of when unsupervised learning has gone wrong. However, the firm RedBalloon was able to get it right. RedBalloon deployed the AI agent "Albert," which was able to analyze customer databases, ad accounts, and ad campaigns through dynamic machine learning proficiencies. Albert immediately eliminated several time-consuming duties that impede the opportunities that digital marketing is known to deliver. Allowing Albert to work unsupervised enabled RedBalloon to achieve unprecedented results that included a 54% reduction in Google Search cost per acquisition (CPA), a 34% increase in Google Search generic ad impressions, a 149% increase in Google Search return on ad spend (ROAS), 1100% return on ad spend, reducing cost per customer acquisition from $50 to less than $17, canceling their $45,000/month retainer with Ad agencies, spending 25% less on marketing, and increasing their overall marketing results by 30% (Sutton, 2018).

Supervised Learning

Supervised learning allows AI-enabled tools to comprehend the data and detect patterns or similarities because it is given labeled input data and the expected output results. An example is the Dali Museum in St. Petersburg, Florida, where they bring Salvador Dali to life through AI. Utilizing over 1000 hrs of machine learning techniques and examining over 6000 frames of historic footage of Dali, the AI-enabled system was able to create 125 different videos that depict the artist's life and an accurately simulated portrayal of his personality. The life-size deep fake can have interactive discussions with Museum patrons and send pictures to their phones after taking selfies.

AI Customer Service Strategy

AI has become a trend that businesses cannot stop talking about. As a cautionary tale, managers may think twice before making such a costly investment—without a well-thought-out AI strategy. It will most likely lead to a siloed project, with the company finding itself at a technological dead-end. Therefore, knowing what AI customer service strategy is might help businesses and organizations figure out whether they should adopt such strategies for optimizing customer services, and if they do, how and in what areas. Simply put, AI customer service strategy helps businesses and organizations *deliver the right solutions to the right target audience—including existing customers and prospects—at the right time and creates an end-to-end memorable customer experience and post-consumption services.*

With the abundance of digital channels, consumers switch between online and offline service channels, leading to increasingly complicated consumer behavior. Therefore, it becomes imperative to equip consumers with the right offers and provide them with the right information. Using technologies such as natural language processing (NLP), Machine Learning (ML), and sentiment and advanced analytics, companies can now gain valuable customer insights by catching sentiments that human agents might overlook, as well as capturing their needs and satisfaction accurately (Dixon, 2018). Such enriched insights then enable companies to feed consumers with more personalized offers, provide relevant helpful information to foster patronage from prospects and deploy intervention strategies to de-escalate consumer complaints and dissatisfaction (Wilson & Daugherty, 2018).

Alternatively, more than stand-alone AI is needed to guarantee a successful AI customer service strategy. It requires companies to implement mindful and strategic maneuvers to deploy human customer service representatives and AI agents to complement and assist each other (Mogaji & Nguyen, 2022). Importantly, a myth in AI applications often centers around how companies view the dynamics among AI, humans, and processes. Companies sometimes use AI to focus on and tackle specific facets of the consumer service journey, including service automation, inventory management, and addressing customer inquiries. However, approaching customer service strategies in isolation entails companies treating both AI and customer service strategies as disconnected individual silos. Such treatment of siloed investment allows for compromised efficiency due to additional labor to understand the situation, wasted processing time, and friction in task delegation and problem-solving. Unfortunately, this can cause consumers to experience frustration as their needs are not being met promptly and concerns are properly. For example, recall the last time you needed to ring a customer service hotline and were referred to an AI chatbot. After an unnecessary time of going back and forth, you ended up hanging up feeling agitated. Indeed, research informs us that although AI outperforms human agents in areas such as mechanical and analytical tasks—for instance, self-service touch-screen kiosks at McDonald's, Pepper the concierge robot, and IBM Watson, it still needs human intervention when it comes to intuitive and empathetic tasks (Huang & Rust, 2018). Without the necessary human touch, consumers often complain that AI-enabled agents, such as self-service chatbots, are unhelpful and merely deployed for cost reduction.

Ideally, when designing a successful AI implementation program, companies may want to embrace the AI–human symbiosis to optimize the outcome of customer service strategies. When AI and humans are strategically deployed across customers' service journeys, there should not be a disconnect where customers are left alone waiting and guessing. Instead, companies should facilitate a frictionless AI–human process to optimize an integrated customer service eco-system that offers holistic assistance and solutions. As a simplified illustration, AI could offer accurate and personalized data-driven insights, leading to more informed personal assistance when compared to human service representatives. In addition, the presence of human services facilitated by AI insight also fertilizes the potential for cross-selling or upselling opportunities (Dixon, 2018; Paschen et al., 2020).

Currently, there is a surge of research in academia on the benefits of AI customer service strategy in a myriad of areas. Whether that might reduce business costs (Kiron & Schrage, 2019), increase customers' patronage due to tailed product/service recommendations, or enhance customer loyalty and stickiness because of hyper-personalized content feeds (e.g., emails, messages, and social media feed). Implemented strategically, AI could supercharge companies' customer service by creating an effortless yet meaningful experience. Nevertheless, approximately 57% of businesses realize that their approach is far from strategic but simply reactive (Zendesk, 2022).

The successful implementation of AI projects goes beyond deploying chatbots as the new customer call center and providing only pre-programmed responses and messages. It requires a mindful search for and selection of data (e.g., *what type of data to look for*), unbiased data analysis, meaningful interpretation (e.g., *what data-driven insights the data can tell us*), and informed decision-making. Before diving into a fruitful endeavor, knowing the prerequisites of a successfully executed AI project is imperative. In what follows, we offer a checklist of pivotal prerequisite factors for companies when designing an AI customer service strategy (Iansiti & Lakhani, 2020).

A Checklist for Implementing Successful AI Projects

- Acquirable and retrievable data
- Organizing structured and unstructured data
- Analyzing data and meaningful interpretation
- Integrating human- and AI-agents to deliver customer-centric services
- Removing (un)conscious biases and discriminations
- Delivering insights into strategic and actionable plans

A Roadmap for Implementing an AI Customer Service Strategy

What constitutes a successful AI customer service strategy? On the surface, this is a complex question. However, technology-driven or not, customer service always comes down to *delivering the right offers to the right target audience –existing customers and prospects—at the right time.* The addition of AI does not and will not change this!

Companies today have no choice but to adopt AI in their customer service strategies. But its addition is not the silver bullet. Companies will have to do it *right.* Looking at AI and what it is capable of allows companies to make sense of customers' needs more effectively by utilizing customer-based data. However, in this process, it is imperative to understand where a customer is along the service journey. With any marketing initiative, knowing what the customer wants is critical.

Customers have different expectations and goals at different stages of their service encounters. To achieve the maximum potential of AI-enabled customer service strategies, companies need to align their customer support with customers' goals and stages of the service journey. By accessing, analyzing, and interpreting customer data using AI, businesses can better fulfill the following goals:

- Segment customers and products into groups that have similar behaviors and needs
- Predict customer purchase and churn risks
- Estimate and optimize the lifetime value of a customer or product
- Increase customer loyalty and encourage repeated purchase
- Address customer concerns accurately and promptly

Ideally, businesses ought to pinpoint customers along their journey, accurately understand a customer's needs, wants, and desires, and utilize AI to deliver services that effectively satisfy them. For instance, when real-time data hints that a customer has been spending an excessive amount of time browsing different service offerings yet failing to patronize, an effective AI customer service system should draw data-driven insights to facilitate purchase behaviors—more resources need to be invested, such as providing personalized recommendations from other like-minded customers (e.g., Netflix recommending new shows to specific customer segments based on similar customer profiles), eliminating options for a customer based on their past experiences, and offering a parallel comparison among viable options.

Similarly, companies can use AI to establish an accurate customer portrait—that estimates and optimizes a customer's lifetime value. Simply put, companies face questions such as: will a customer return or be gone for good? How much profit can the company make from a particular customer in the future? Gathering a customer's past transactional data helps answer these questions. Fitting customers to a probabilistic model using historical and predictive customer lifetime value analysis redirects companies' focus to questions such as:

- How do you prioritize resources depending on a customer's lifetime value?
- How do you identify prospects and convert them into customers?
- How do you identify potentially abandoned carts from customers beforehand and take preventative measures?
- How do you utilize customer relationship management for customer retention?

Furthermore, effective AI customer service strategies also empower businesses to engage in proactive support—*knowing and offering your customers' needs before they ask for them.* By identifying and addressing issues even before customers realize them, AI can optimize customers' service journeys seamlessly, leading to a memorable end-to-end experience. Therefore, a successful AI project goes beyond the computation of big data. Its supremacy can only be achieved in a well-thought-through intelligence eco-system—the ability to capture, organize, analyze, and interpret consumer data at scale; the utilization of AI to understand consumer behavioral patterns; the deployment of AI–human symbiosis to shape, personalize, and optimize a customer's journey.

We next offer a roadmap that portrays what successful AI customer service strategies might look like. By mapping out different stages in customers' journey, it allows businesses to supercharge their customer service strategy—surgically understanding customers' goals and pain points, minimizing friction by aligning the best customer service personnel (AI, humans, or both) with customer needs, and offering personalized service solutions in real-time (see Fig. 4.1).

Step 1: Triage

Finding the Root Causes of Problems, Accurate Re-Directions, and Task Delegation

In an era where consumers often swap between online and offline channels, not to mention a myriad of different digital channels, we are at the point where competitive advantage derives from knowing exactly where customers are along their journey and offering them exactly what they need. For companies who are failing to do so, we cannot necessarily blame them for their lack of trying. Compared to humans, AI can capture subtle sentiments hidden in qualitative data, such as social media footprints and their interactions with companies—insights otherwise missed or ignored.

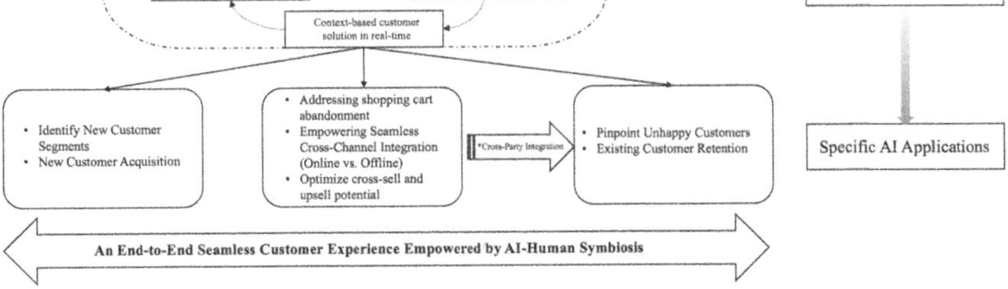

Fig. 4.1 AI customer service strategy roadmap *Note 1*. Figure developed by the authors. *Note 2*. Cross-party integration occurs when customers expect support from multiple platforms, such as Facebook, Snapchat, TikTok, etc.

When picturing AI and humans working together, people might spontaneously assume that humans will give instructions while AI "listens to" such instructions and executes the commands. However, AI has climbed up the organizational pyramid and encroached on critical management tasks, such as task delegation.

AI can surgically pinpoint customers along their journey by picking up specific vocabularies during companies' interactions with customers. For example, using linguistics-based Natural Language Processing (NLP), keywords such as "repairs" or "issue" and "disappointed" are more likely to be collated with post-consumption complaints instead of pre-consumption inquiries. Companies equipped with these data-driven insights can then redirect the customer to the department of customer service, complaints, or warranty and decide whether to interject to de-escalate the situation accordingly.

Real-World Examples: VirtuOz and WorkFusion

Consider VirtuOz, a San Francisco-based virtual agent company. VirtuOz is one of the leading pioneers in using AI to analyze conversations with customers on the phone, figure out what the customers' intentions are, and find the right human agent to connect the customer to without the customer explicitly asking. Now, VirtuOz represents companies using AI to delegate tasks to humans. Then there's WorkFusion. Unlike VirtuOz, WorkFusion uses Robotic Process Automation (RPA), intelligent conversational agents, and analytics to process the workflow required and breaks many complicated problems into discrete tasks allocated to either AI or human agents (Barr, 2015).

Step 2: Re-imagine a Seamless End-to-End Experience

After a preliminary task delegation based on AI's data-driven insights, a customer's needs can be categorized into one of three domains: prospects, pre-consumption, and post-consumption. All streams lead to the same pond—the key to customer service is converting prospects into customers, eventually to repeat customers. So, let's embark on the journey of revisiting an end-to-end customer service experience through the lens of AI—from finding and acquiring prospects to optimizing the pre- and mid-consumption experience and elevating a painless post-consumption experience (Mogaji et al., 2020). In what follows, let's dive deeper into each stage and uncover how AI helps.

Identifying and Acquiring Prospects

Leveraging AI-enabled technologies, an AI customer service strategy ensures that every piece of marketing communication is tailored to the prospects' liking, leading to substantially higher conversion rates from prospects to customers.

From AI social influencers to AI-enabled content marketing, intelligent opportunities offered by AI have helped businesses identify prospects, optimize leads, and convert prospects into customers (Mazzini, 2020). Think of Facebook. You might have followed brands you are interested in and "liked" pages that grabbed your attention. These behaviors do not go unnoticed. Companies consider your interests when they attempt to reach out to you. They might communicate with you about an upcoming concert if you frequently browse and respond to music-related content.

AI is transforming the content creation spheres every second of every day. AI-empowered content can elevate businesses' customer service by accelerating the content creation process, crafting content that is tailored to customers' search intent, and delivering hyper-personalized messages to targeted audiences based on their liking and preference. For example, RankBrain, an algorithm developed by Google, analyzes customers' search keywords and user behaviors to discover what they want when they search online (Boyd, 2018). AI-enabled tools like RankBrain allow businesses to create content that might appeal to them and thus increase their search presence. Taking it one step further, AI is powerful in sending targeted, personalized content to grab consumers' attention and stimulate responses. For instance, Personyze, an AI-enabled personalization engine, tracks customers' preferences and establishes a visitor profile using their social media presence, demographics, location, and referral. Subsequently, each customer receives personalized content tailored to them (Authority Magazine, 2021). After delivering content to the target audience, AI also helps monitor the performance of such content. Take Zia as an example, a tech-savvy company that collects customer-based data about marketing leads and makes suggestions on which leads are most likely to close, even reminding the business of the best time of a day to follow up on a lead with a prospect.

With the rise of the audience base using social media, it becomes increasingly challenging for brands to stand out in the crowd. Using AI, some brands are now creating their social influencers. Virtual influencers created using AI offer brands more control over their content creation than human social influencers. More importantly, CGI-based creations are designed to cater to target audiences more effectively. Specifically, by analyzing the prospects, virtual influencers assemble an accurate embodiment of qualities and characteristics that the target audience finds attractive and can relate to. Subsequently, they can effectively engage with the prospects, converting them into customers. For example, Lil Miquela, one of the most popular virtual influencers, has over three million followers on Instagram. Lil Miquela, portrayed as a 19-year-old figure living in Los Angeles, has been featured in fashion magazines and as the ambassador for a range of luxury brands, such as Prada, Dior, and Calvin Klein (Rabimov, 2020).

Real-World Example: Harley-Davidson

Meet Albert! Selling motorcycles in the fall and winter seasons can be challenging. When Harley-Davidson was selling only one or two motorcycles per week, this needed to be better, thought Asaf Jacob, the president of Harley-Davidson NYC.

They needed to do something different. Traditional marketing that targets everyone does not work for them. To provide better customer service, this iconic motorcycle brand turned to AI for help. In 2016, Harley-Davidson launched its first AI marketing platform, Albert. In the first two days after the campaign's launch, Albert has sold over 15 bikes for Harley-Davidson, compared to pre-AI average sales—1–2 motorcycles per week.

In the long term, the creative use of AI has increased their New York sales leads by 2930% per month (Power, 2017).

So, how does Albert work? By analyzing previous customer data, Albert can identify key determining characteristics of their existing high-value buyers, such as age, gender, location, etc. The buyer persona identified, therefore, is used to find prospects who share similar profiles as their high-value buyers, or according to Harley-Davidson, the "lookalikes." Due to the demographic and behavioral similarities shared among the existing high-value buyers and the "lookalikes," Albert helped Harley-Davidson to isolate the niche target customer segment who are more likely to make a purchase, which only accounted for approximately 2% of the population (Marr, 2018).

Here's an example where Harley-Davidson puts Albert to work. By analyzing the browsing data on the website, Albert categorizes prospects into different sub-segments—those who can be converted and those who cannot—then approaches these segments strategically. For instance, during a service interaction, a visitor who spends less than 15 seconds will be dismissed by Albert as they are unlikely to be converted. In contrast, Albert will isolate visitors who spend roughly one to two minutes on the site of Harley-Davidson and start following up with them. In addition, the behavioral patterns of existing customers also help Harley-Davidson uncover new customer segments they conventionally overlook. Albert found that people who purchase high-end jewelry such as diamonds are also likely to purchase motorcycles. Therefore, this seemingly random association helped Harley-Davidson break through—a missing customer segment typically ignored by motorcycle advertisements (Power, 2017).

Optimizing the Pre- and Mid-consumption Experience

As AI continues to evolve, along with supporting technologies such as machine learning (ML) and access to big data, companies have been integrating AI into their operations to remove friction points and improve customer service experience.

Capitalizing on historical analysis based on past and current customer data can shed light on if a customer will make a purchase (whom), what types of products (what), when they might purchase (when), and through what channels (how). This is called predictive analytics. Predictive analytics predominantly uses data mining, statistical modeling, and machine learning (Christoffersen, 2017). Through the process of making data-inspired predictions, it has been leveraged to optimize the conversion from prospects to customers and eventually to repeat customers—for example, it can predict if a customer is likely to (re)purchase products, what they will (re)purchase and when (Grossi et al., 2022).

Not every prospect will make a purchase. If your company has witnessed considerable website traffic yet failed to close deals, one possibility might be that they are abandoning their shopping cart. According to a recent study, the shopping cart abandonment rate is at a staggering 70%. Therefore, looking into why people abandon their carts ought to help businesses understand what makes customers purchase or not purchase and eventually boost sales.

Customers might decide not to purchase for many reasons: encountering payment issues, excessively high additional costs (e.g., shipping, tax, service surcharge), difficult checkout process, or simply because they haven't decided yet. Because of AI, companies can surgically pinpoint why an individual customer decides not to proceed with their purchase. For example, for cost-conscious customers, AI-empowered chatbots can nudge customers to complete the purchase by showing them limited-time coupons or free-shipping codes. In addition, despite privacy concerns, consumers still prioritize personalized experiences. When customers haven't decided to purchase what's inside their online baskets, AI can glean customer data, such as browsing history, previous purchases, and social media feed, and make hyper-personalized and tailored product/service recommendations.

Used correctly, predictive analysis embedded in AI can be adopted to promote additional purchases among existing customers. This is a win-win situation for marketers, who do so effectively—optimizing their existing customers' lifetime cycle by meeting customers' needs and boosting their sales without having to acquire new customers. Leveraging algorithms (e.g., the Apriori algorithm), AI can conduct market basket analysis (MBA) and establish product associations to identify the optimal combinations of products frequently bought together (Christoffersen, 2017).

Companies can then utilize the insights of the analysis to promote more cross-sell and upsell actions. Moreover, with the marketplace becoming increasingly competitive, every business—big or small—needs to navigate through the dilemma of retaining existing customers and encouraging repeat purchases. With predictive analytics, businesses can better allocate resources to stimulate repeat customers. By focusing on different customer segments strategically, customer retention can be made easier. For customers who are less loyal to your company, using highly targeted and personalized messages can motivate them to return and repurchase.

Real-World Example: Netflix

For a business that kicked off offering DVD rental services, Netflix is doing well! Especially, since the COVID-19 pandemic, it is expected that Netflix will present a compound annual growth rate (CAGR) of 19% and a market value of US$438.5 billion. So, what is the recipe? Well, embracing AI might be the secret ingredient.

With thousands of shows and movies, customers will only invest a few minutes before they give up. Using machine learning and big data, Netflix offers a novel solution—providing each customer with movie recommendations that they most likely resonate with. Customers who watch movie A are more likely to watch movie B. This mantra helps Netflix to use watching history and recommend movies of a similar genre to customers with a compatible user profile.

Netflix has taken personalization another step further. Leveraging AVA technology, algorithms, facial recognition, and object detection, Netflix developed a unique solution to attract customers' attention—feeding them with movie thumbnails that might interest them. If a customer likes and frequently watches romantic comedies, creating a movie thumbnail that highlights the romance genre or emphasizes prominent Rom-Com casts—for example, Matthew McConaughey—might lead to significantly higher viewership (Cold Call, 2022).

Elevating a Painless Post-consumption Experience

The holy grail of customer service at this stage must be proactive and predictive! Analyzing previous customer data allows companies to know what customers need before they know it.

Why should businesses care about one's post-consumption experience? On average, acquiring a new customer costs five times as much compared to retaining an existing one. Not all customers are equal, and customer retention is a cost-effective strategic move. However, knowing how customers feel about their service experiences can be challenging, especially for small to medium businesses with limited resources to filter and access such data.

This is where AI can help. Leveraging technologies such as machine learning and deep learning, AI allows businesses to identify "happy" and "unhappy" customers as it can process a massive amount of data, filter relevant information such as feelings and emotions, gather both textual, voice-related, facial, and bodily cues, and redirect companies' focus to retain customers more effectively.

Consider sentiment analysis in making sense of customers' online reviews—whether on the company website or their social media page. Sentiment analysis using AI works toward the same goal as focus groups and in-depth interviews, providing insights with greater accuracy without sacrificing time, money, and other expenses. Scanning customers' comments, feedback, and other forms of Word of Mouth (WOMs)—good and bad—helps companies identify "happy" and, more importantly, "unhappy" customers in real-time. That way, you promptly know the severity of customer dissatisfaction and when to adopt intervention tactics by reaching out to that customer and offering mitigation plans.

In addition to analyzing textual data, the growth of virtual personal assistants (VPAs) allows customers to interact with AI through voice. According to Globe Newswire, by 2025, the speech and voice recognition market will be approximately $26.8 billion (OTO systems Inc., 2020). AI implementation, such as sentiment analysis, can also help businesses understand how well their customer support team performs. For example, it can pick up subtle cues such as intonation, accent, and other voice-related indicators by capitalizing on machine learning and speech detection. Doing so sheds light upon human agents' tone when answering customers' calls, such as boredom, stress, and satisfaction. Subsequently, companies can determine effective phone calls with customers and the areas where their teams may require improvement and training.

Real-World Example: Affectiva

Affectiva is a company that develops AI systems to comprehend human emotions via visual cues—facial expressions, body language, gestures, and auditory cues—the tone of their voices. Instead of asking customers to self-report their satisfaction by clicking one of the given (un)smiley emojis, which can be wildly unreliable, Affectiva provides customer support at a higher level.

Companies that adopt a reactive approach often attempt to gauge customer satisfaction by risking annoying customers by clicking buttons and asking endless questions, then react accordingly. Challenging such traditions, Dr Taniya Mishra, the director of AI research and lead speech scientist at Affectiva, approaches customer service more proactively. In their mind, it is the companies' job to gauge customers' moods and offer solutions. And what can work better than seeing the customers and hearing the tone of their voices. This is when Affectiva steps in. Simulating how humans gather information through communication, Affectiva's AI-enabled customer service fully capitalizes on customers' visual presence and experiences rather than just their responses on a Likert scale and Emoji. By analyzing their biometric and behavioral data in a virtual environment, Affectiva empowers companies to glean and collect more data that customers might consciously or unconsciously mask. The more businesses know about their customers, the better customer service can be provided.

Implications

Technology is changing how the game of customer service is played. This is especially true as businesses navigate massive amounts of customer-based data and the ever-changing customer expectation landscape. This compels us to take a fresh look at AI customer service strategy and what that means for academia, industry, and society.

Academia

While most of the spotlight in academia has been focused on the functional and mechanical development of AI, it remains *an unconquered territory* regarding how customers respond to AI customer service. Research informs us that some customers remain skeptical about AI customer service (Dietvorst et al., 2015), whereas others welcome it in the marketplace (Logg et al., 2019). Academics, therefore, ought to benefit by investigating how deploying AI shapes customers' perceptions, expectations, and attitudes toward AI-enabled customer service.

Industry

Companies that fail to embrace the AI revolution strategically could get left behind. For example, more than 70% of customers expect service encounters to be conversational and feel human, yet only 40% of businesses have managed to deliver (Zendesk, 2022). Indeed, despite AI's undeniably superior advantages in task delegation and data processing, it still requires human intervention, even instructions sometimes, especially for tasks that demand empathy, emotional intelligence, and other complicated decision-making skills. Therefore, customers still expect service experiences to have a "human touch," which compels businesses to foster an AI–human symbiosis rather than replacing human agents with AI for good. More importantly, the successful integration of AI and human agents ultimately should lead to a seamless and wait-free handover between AI and human agents.

Given that investing in AI is not the sole answer, businesses must determine the following questions before they dive into the AI adventure: What does AI mean for our business? How can AI optimize our customer service? What areas does our business still require human talent?

Society

As AI continues to penetrate our lives, its societal implication becomes worthy of more discussion. AI is here to augment instead of replacing human intelligence. A mindful deployment of AI is thus built upon full consideration of human well-being: employment, sense of meaningfulness at work, sense of belonging, and connectedness.

Moreover, has AI been applied ethically? AI has fundamentally revolutionized every step of the customer journey. Are there any vulnerable customers being neglected? Rejected? Or abused? AI operates on the back of massive amounts of customer-based data. However, data processing, training and analysis might be biased against minorities and ill-represented customer segments (Burt, 2021). Algorithms applied in loan processing might discriminate between customers with different demographic backgrounds; in the pharmaceutical industry, it has been reported that, when handling requests for drug prescriptions, AI discriminates against low-income customers (Campbell et al., 2020). Therefore, society must generally reflect on the ethical bottom line in applying AI in customer service.

Conclusion

A customer service strategy is an essential recipe for business success. AI has drastically revolutionized the business playground from identifying, acquiring, and retaining customers. Being a powerful helper, AI only sometimes does things differently from human agents. It does the same thing(s) on most occasions—it does it better!

To realize the full potential of an AI customer service strategy, AI and human agents should work hand in hand, and then voila! Establishing the optimal orchestration not only allows businesses to cut costs and enhance workflow and productivity but also creates an end-to-end memorable customer experience, from identifying and acquiring prospects to optimizing pre- and mid-consumption experience and eventually delivering a painless post-consumption experience. The AI–human symbiosis connects data signals to data-driven insights, engages in AI-human customer service integration, and offers context-based customer solutions in real time. And the rule of engagement? Find the right customer service agents—whether AI, humans, or a combination of both—to deliver a frictionless, personalized, strategic, and 24/7 customer experience!

References

Anantrasirichai, N., & Bull, D. (2021). Artificial intelligence in the creative industries: A review. *Artificial Intelligence Review, 55*(1), 589–656.

Authority Magazine. (2021). Amra Beganovich on how to use digital transformation to take your company to the next level. *Medium*. https://medium.com/authority-magazine/amra-beganovich-on-how-to-use-digital-transformation-to-take-your-company-to-the-next-level-71efff92e053

Barr, K. (2015). AI is getting good enough to delegate the work it can't do. *Harvard Business Review*. https://hbr.org/2015/05/ai-is-getting-good-enough-to-delegate-the-work-it-cant-do?registration=success.

Borges, A. F., Laurindo, F. J., Spínola, M. M., Gonçalves, R. F., & Mattos, C. A. (2021). The strategic use of artificial intelligence in the digital era: Systematic literature review and future research directions. *International Journal of Information Management, 57*, 102225.

Boyd, C. (2018). If a computer wrote this, how would you know. *Medium*. https://medium.com/swlh/if-a-computer-wrote-this-how-would-you-know-the-story-of-ai-and-content-creation-7a597d6ace2a.

Burt, A. (2021). How to fight discrimination in AI. *Harvard Business Review*. https://hbr.org/2020/08/how-to-fight-discrimination-in-ai.

Campbell, C., Sands, S., Ferraro, C., Tsao, H. Y. J., & Mavrommatis, A. (2020). From data to action: How marketers can leverage AI. *Business Horizons, 63*(2), 227–243.

Christoffersen, S. (2017). *6 ways to use data science to drive your cross-sell and upsell activity*. IBM. https://www.ibm.com/blogs/nordic-msp/six-ways-use-data-science-drive-cross-sell-upsell-activity/.

Chung, K., Thaichon, P., & Quach, S. (2022). Types of artificial intelligence (AI) in marketing management. In *Artificial intelligence for marketing management* (pp. 29–40). Routledge.

Cold Call. (2022). Delivering a personalised shopping experience with AI. *Harvard Business Review*. https://hbr.org/podcast/2022/05/delivering-a-personalized-shopping-experience-with-ai

Cui, Y., & van Esch, P. (2022). Autonomy and control: How political ideology shapes the use of artificial intelligence. *Psychology & Marketing, 39*(6), 1218–1229.

Cui, Y. G., van Esch, P., & Jain, S. P. (2021). Just walk out: The effect of AI-enabled checkouts. *European Journal of Marketing*. https://doi.org/10.1108/EJM-02-2020-0122

Dastin, J. (2018). Amazon scraps secret AI recruiting tool that showed bias against women. In *Ethics of data and analytics* (pp. 296–299). Auerbach Publications.

Dietvorst, B. J., Simmons, J. P., & Massey, C. (2015). Algorithm aversion: People erroneously avoid algorithms after seeing them err. *Journal of Experimental Psychology: General, 144*(1), 114.

Dixon, M. (2018). Reinventing customer service. *Harvard Business Review, 96*(6), 82–90.

Grossi, D., Van Der Hoek, W., & Kuijer, L. B. (2022). Reasoning about general preference relations. *Artificial Intelligence, 313*, 103793.

Guttmann, A. (2022). Leading areas combining AI and marketing automation worldwide in 2022. *Stat*. https://www.statista.com/statistics/1303223/top-areas-artificial-intelligence-marketing-automation/.

Huang, M. H., & Rust, R. T. (2018). Artificial intelligence in service. *Journal of Service Research, 21*(2), 155–172.

Iansiti, M., & Lakhani, K. R. (2020). *Competing in the age of AI: Strategy and leadership when algorithms and networks run the world*. Harvard Business Press.

Kiron, D., & Schrage, M. (2019). Strategy for and with AI. *MIT Sloan Management Review, 60*(4), 30–35.

Logg, J. M., Minson, J. A., & Moore, D. A. (2019). Algorithm appreciation: People prefer algorithmic to human judgment. *Organizational Behavior and Human Decision Processes, 151*, 90–103.

Marr, B. (2018). How Harley-Davidson is using AI and robots to prepare for the 4th industrial revolution. *Harvard Business Review.* https://www.forbes.com/sites/bernardmarr/2018/07/03/how-harley-davidson-is-using-ai-and-robots-to-prepare-for-the-4th-industrial-revolution/.

Mazzini, C. (2020). Five ways AI is changing the game of content marketing. *Forbes* https://www.forbes.com/sites/forbesagencycouncil/2020/03/20/five-ways-ai-is-changing-the-game-of-content-marketing/?sh=15810e136559.

Mogaji, E., & Nguyen, P. N. (2022). Managers' understanding of artificial intelligence in relation to marketing financial services: Insights from a cross-country study. *International Journal of Bank Marketing, 40*(6), 1272–1298.

Mogaji, E., Soetan, T., & Kieu, T. (2020). The implications of artificial intelligence on the digital marketing of financial services to vulnerable customers. *Australasian Marketing Journal., 29*(3), 235–242.

Montes, G. A., & Goertzel, B. (2019). Distributed, decentralized, and democratised artificial intelligence. *Technological Forecasting and Social Change, 141*, 354–358.

OTO systems Inc. (2020). Voice AI: From personal assistants and beyond. *Medium.* https://medium.com/voice-tech-podcast/voice-ai-from-personal-assistants-and-beyond-350bb9edfc2c.

Paschen, J., Wilson, M., & Ferreira, J. J. (2020). Collaborative intelligence: How human and artificial intelligence create value along the B2B sales funnel. *Business Horizons, 63*(3), 403–414.

Pitt, C., Paschen, J., Kietzmann, J., Pitt, L. F., & Pala, E. (2021). Artificial intelligence, marketing, and the history of technology: Kranzberg's Laws as a conceptual lens. *Australasian Marketing Journal, 31*(1), 81–89.

Power, B. (2017). How Harley-Davidson used artificial intelligence to increase New York sales leads by 2,930%. *Harvard Business Review, 30*. https://hbr.org/2017/05/how-harley-davidson-used-predictive-analytics-to-increase-new-york-sales-leads-by-2930

Rabimov, S. (2020). The real game-changers: How virtual models take on creative industries. *Forbes*. https://www.forbes.com/sites/stephanrabimov/2020/06/10/the-real-game-changers-how-virtual-models-take-on-creative-industries/.

Sahota, N., & Ashley, M. (2019). *Own the AI revolution: Unlock your artificial intelligence strategy to disrupt your competition: Unlock your artificial intelligence strategy to disrupt your competition*. McGraw Hill Professional.

Sopadjieva, E., Dholakia, U. M., & Benjamin, B. (2017). A study of 46,000 shoppers shows that omnichannel retailing works. *Harvard Business Review, 3*, 1–2. https://hbr.org/2017/01/a-study-of-46000-shoppers-shows-that-omnichannel-retailing-works

Sutton, D. (2018). How AI helped one retailer reach new customers. *Harvard Business Review*. https://hbr.org/2018/05/how-ai-helped-one-retailer-reach-new-customers.

van Esch, P., & Cui, Y. (2021). Does consumer promiscuity influence purchase intent? The role of artificial intelligence (AI), change seeking, and pride. *Journal of the Association for Consumer Research, 6*(3), 394–401.

van Esch, P., Cui, Y., & Jain, S. P. (2021). Self-efficacy and callousness in consumer judgments of AI-enabled checkouts. *Psychology & Marketing, 38*(7), 1081–1100.

Wilson, H. J., & Daugherty, P. R. (2018). Collaborative intelligence: Humans and AI are joining forces. *Harvard Business Review, 96*(4), 114–123.

Xu, Y., Shieh, C. H., van Esch, P., & Ling, I. L. (2020). AI customer service: Task complexity, problem-solving ability, and usage intention. *Australasian Marketing Journal (AMJ), 28*(4), 189–199.

Zendesk. (2022). *Unlocking growth with emerging capabilities*. https://www.zendesk.com/cx-trends-ai-conversational/

Theme II

Customer Engagement

5

Personalized and Contextual Artificial Intelligence-Based Services Experience

Valentina Pitardi

Introduction

Nowadays, Artificial Intelligence (AI) service applications in the form of recommendation systems, voice-activated assistants, chatbots, and embodied robots, are becoming increasingly used in the marketplace and are expected to replace human service providers in numerous industries (De Keyser & Kunz, 2022). This pervasive use of AI-based services offers novel opportunities for companies that want to enhance the customer experience and provide customers with timely, personalized services. For example, individuals can engage with AI applications and chatbots in various medical settings, where the technology is used to diagnose and dispense medical advice to consumers (Longoni et al., 2019). Customers can find service robots such as Nao and Pepper serving their tables or giving them information on several hospitality and tourism frontlines

V. Pitardi (✉)
Surrey Business School, Guildford, UK
e-mail: v.pitardi@surrey.ac.uk

© The Author(s), under exclusive license to Springer Nature Switzerland AG 2023

J. N. Sheth et al. (eds.), *Artificial Intelligence in Customer Service*,
https://doi.org/10.1007/978-3-031-33898-4_5

(Mende et al., 2019). Almost every person with a bank account has received support from a virtual chatbot during online customer service (Hildebrand & Bergner, 2021).

However, the numerous possibilities that people have today to experience such services make it difficult for companies to identify the conditions under which AI-driven services may improve rather than worsen the overall customer experience. Research has also emphasized that if AI can lead to positive evaluations and outcomes (Logg et al., 2019), there are situations where the same interactions may result in dissatisfaction, disengagement, and negative feelings (Longoni et al., 2019).

What complicates the picture is that customers' interactions with AI are not limited to applications embedded in brands' platforms. Many people already interact with AI agents and services in everyday life, for instance, when using their Google Home speaker or Fitbit's fitness tracker. These daily interactions shape the customer experience with AI and contribute to rising expectations toward the technology, posing a challenge for companies that aim to deliver effective and successful services.

This chapter aims to bring clarity to the topic. Building on recent AI and consumer behavior literature offers a framework to help companies identify when and how AI-based services can enhance the customer experience through personalized services.

AI and Customer Experience

Artificial Intelligent Technology: A Definition

Studies from computer science refer to AI as a set of programs and systems able to imitate human behavior and perform tasks normally requiring human intelligence, such as problem-solving and autonomous decision-making (Russell & Norvig, 2009). AI can then be embedded in various technologies that serve different purposes (e.g., search engines, virtual assistants, algorithmic trading systems, etc.). In a nutshell, AI, like electricity, is a general-purpose technology that can have a variety of applications.

While AI indicates the core technology, in the marketing literature, the name is used as an umbrella term to describe a multitude of applications, including algorithmic recommendation systems (e.g., Netflix), voice assistants (e.g., Amazon Alexa), chatbots (e.g., Cleo), and robots (e.g., Pepper). As a result, the definitions available in the literature vary based on whether they focus on the core functioning of the technology (i.e., the set of programs), its application (i.e., search engines, recommendation systems), or both. For example, some scholars have adopted a more technical approach and defined AI as "a system's ability to interpret external data correctly, to learn from such data, and to use those learnings to achieve specific goals and tasks through flexible adaptation" (Kaplan & Haenlein, 2019, p. 17). Most service research has focused on using this technology for service provision and customer interface and defined AI agents as "system-based autonomous and adaptable interfaces that interact, communicate and deliver service to an organization's customers" (Wirtz et al., 2018, p. 909). Finally, research on consumer behavior and psychology has taken a more general approach and defined AI as "any machine that uses any kind of algorithms or statistical model to perform functions typical of the human mind" (Longoni et al., 2019, p. 2).

All these definitions emphasize three fundamental capabilities that distinguish AI from the previous type of technology. First, AI-based technology can *learn* from data and *adapt* autonomously. This capability allows AI agents to offer more flexible interactions and service scripts and recover common service failures (Choi et al., 2021; Mogaji et al., 2020). Second, AI can *interpret* and *predict*, which allows these applications to offer ultra-customized offerings and maximize satisfaction (Abdulquadri et al., 2021; Kumar et al., 2019). Third, AI-based technologies can *interact* and *communicate* with users by mimicking human-like conversations. The ability to engage in such social interactions can result in more positive consumer evaluations (Belanche et al., 2021; Yoganathan et al., 2021), increased trust (van Pinxteren et al., 2019), and customer engagement (Huang & Rust, 2021).

These capabilities allow AI agents to perform different types of tasks requiring various levels of intelligence (i.e., mechanical, thinking, and feeling intelligence; Huang & Rust, 2021). The level of personalization that AI technology can deliver will depend on that intelligence. To date,

AI agents are mostly deployed in transactional services. They have become particularly successful in performing mechanical tasks (e.g., routine and repetitive tasks) and thinking intelligence (e.g., large amounts of data processing and personalized outputs). For example, Pepper robots are used in hospitality services to greet customers and help them navigate the service scape. At the same time, the chatbot Cleo provides customers with complex financial auditing and analyses. Given the focus on repetitive tasks, mechanical AI offers the lowest level of personalization. Conversely, thinking AI can collect and analyze large amounts of contextual consumer data, allowing managers to offer customers hyper-personalized services and interactions (Mogaji & Nguyen, 2022). For example, thinking AI supports search engines like Google in providing personalized search results and recommendations. Insurify, an American company, uses AI to offer its users real-time personalized home and care insurance planning that match users' specific needs.

As for feeling intelligence, the adoption of AI agents for more relational services is still in its infancy, even if the recent diffusion of Replika—a virtual AI companion and friend—already provides some insights into the use of the technology for such types of tasks (Liu-Thompkins et al., 2022). Moreover, existing AI applications such as voice assistants (i.e., Amazon Alexa or Apple Siri) already provide some forms of relational interactions with customers. However, feeling AI is not yet fully developed, and so is the potential personalization that this type of AI could help achieve.

AI, Customer Journey, and Personalized Experience

When embedded throughout the customer journey, AI-based technologies can enhance the overall customer experience bringing benefits to companies. However, the ways AI could impact the customer experience will depend on the capabilities of the technology and the stage of the customer's journey (Daqar & Smoudy, 2019).

In the pre-purchase stage, the customer experience may be enhanced by the ability of AI to learn, adapt, and predict. In this stage, the proactivity of AI technology can help direct customers toward more suitable and profitable solutions. At the same time, predicted preferences and personalized recommendations can help customers navigate the pre-purchase stage faster and more efficiently. For example, beauty brands like Sephora use AI virtual agents to help customers identify the most appropriate skincare, haircare, or makeup.

During the purchase stage, the experience can be improved by AI-personalized recommendations and the ability to interact with users. By learning about customer preferences, AI can provide alternative and complementary products and service recommendations that match individuals' interests. Amazon's "you also look at this item" or "other people also bought this" types of recommendations are generally based on individuals' past purchases or rated items and are an example of this application.

In the post-purchase stage, the capability to interact and communicate with users plays the most relevant role. AI agents can engage customers in a two-way conversation, providing support across platforms and devices. Moreover, they can identify potential signs of dissatisfaction in advance, providing feedback and solutions to take the appropriate actions toward the customers. Virtual agents such as Jamie of ANZ bank are examples of such applications.

When focusing on the customer experience, the technology's design and configuration is a relevant aspects that might influence how individuals respond and react to AI-based services. Consumers, for instance, interact with AI services when buying items on Amazon, when talking to Siri or Cortana, or when chatting with a support chatbot. Suppose some recommendation systems appear as 'simple' websites (i.e., Amazon). In that case, most AI applications today materialize in various types of agents that differ by modality, tangibility, and appearance (Wirtz et al., 2018). In this sense, AI applications can vary based on their mode of interaction (e.g., text, voice, and touch). For example, Amazon Alexa or Apple Siri allow for voice interactions, while most virtual chatbots only support text-based interactions. Second, AI agents can be virtual (e.g., Replika, AI virtual friend) or embodied (e.g., Pepper). Third, their appearance can vary from machine-like (e.g., baggage handling robots in hotels) to human-like (e.g., Nao) to humanoid (e.g., Sophia).

The mix of such characteristics and the final configuration of the AI application can greatly impact the customers' general acceptance of the technology and experience with it. For example, humanoid robots appear to generate more discomfort and uneasiness in customers than their machine-like counterparts (e.g., Mende et al., 2019). However, there is no magic formula, and the same configuration can improve the effectiveness of the service for specific tasks and types of consumers while worsening it for other functions and segments. Thus, it is becoming increasingly difficult for practitioners to understand when AI personalized services enrich the customer experience and maximize engagement and satisfaction.

When and How AI Personalized Services Enhance the Customer Experience

AI-Based Services: Avoidance and Acceptance

There is burgeoning research investigating customers' usage and acceptance of AI-based services (e.g., Huang & Rust, 2018, 2021; van Doorn et al., 2017; Wirtz et al., 2018). However, empirical evidence is still mixed on whether individuals accept or resist AI-based services.

On the one hand, people tend to resist AI-personalized services and recommendations. This phenomenon is called 'algorithm aversion' (Dietvorst et al., 2015). It has been largely documented in the literature where studies showed that customers prefer human-based services in various AI applications, including medical decision-making (e.g., Longoni et al., 2019), recruitment (Lee, 2018), finance services (Önkal et al., 2009), and content suggestions (Yeomans et al., 2019). There are numerous reasons why people resist AI-based services. For example, people rely less on AI-based medical services because they believe that the service and care provided by the AI will not consider their unique individual conditions (Longoni et al., 2019). Similarly, individuals rely less on AI-based suggestions when forecasting stock prices because AI technologies cannot explain their decisions (Önkal et al., 2009). When used by the organization in selection decisions, AI-based services are generally considered less fair and less trustworthy than human decision-makers (Lee, 2018). Moreover, individuals believe that AI technologies cannot learn and improve and thus trust them less when they see them making mistakes (Dietvorst et al., 2015).

On the other hand, a growing number of studies have observed the opposite phenomenon and supported what is called 'algorithm appreciation.' As proposed by Logg et al. (2019), this stream of research shows that humans prefer and adhere more to AI-based services than humans. For example, it appears that people rely more on AI-based services when they believe that the AI has greater expertise than themselves regarding the service matter (Banker & Khetani, 2019), when the AI messages highlight 'how' action should be performed, rather than 'why' (Kim & Duhachek, 2020) or when they feel embarrassed (Pitardi et al., 2021).

Research on AI aversion and AI appreciation provides contrasting answers to whether consumers favor AI or human-based services, leaving open challenges for companies that want to identify the best strategies and techniques to implement the technology. However, when looking at how customers interact with AI-based service, such a dichotomous approach that positions customers either on the avoidance or the acceptance extreme appears rather artificial. It is very rare to find people steadily accepting or resisting AI services in the reality of consumption. Building on this, we focus on the overall spectrum of the AI customer experience in the next section. We propose a three-way perspective that can help map when and how AI-based services can improve customer experience through personalization.

A Three-Way Perspective on the AI-Based Service Customer Experience

Research has emphasized that customers experience AI differently based on contextual, personal, and technological characteristics and showed that, when situational, such elements can turn individuals' avoidance into acceptance. In this section, we integrate these perspectives and propose a three-way framework to interpret and evaluate the AI-based

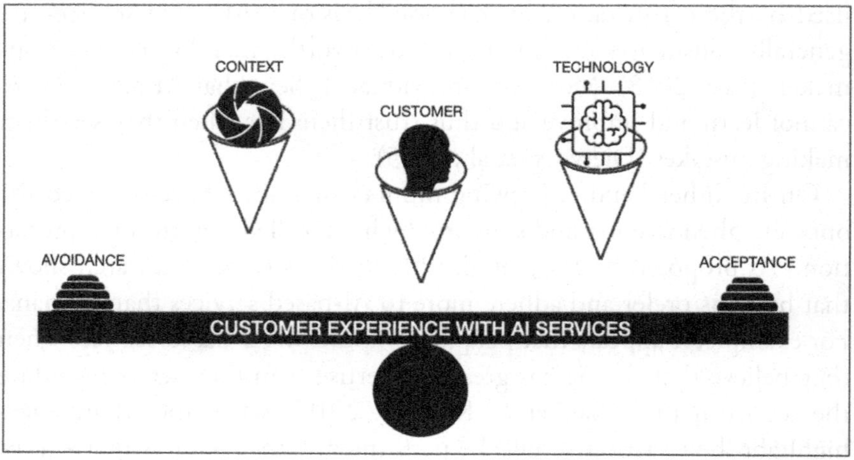

Fig. 5.1 Three-way perspective on customer experience with AI-based services

service customer experience (see Fig. 5.1). Each dimension—the context, the customer, and the technology—identifies relevant characteristics that can affect the overall AI-service experience. Identifying these features is relevant and provides companies with practical and actionable suggestions on the conditions where AI-based services are accepted and preferred to humans.

The Context: Product, Task, and Encounters

Generally, the consumption contexts, including tasks' attributes and product types, influence consumers' acceptance and experience of AI-based personalized services.

First, when the application context is *utilitarian*, AI-based personalized recommendations are trusted and accepted more (Longoni & Cian, 2022). Utilitarian consumption generally refers to products and services that provide customers with instrumental, functional, or practical utilities. In these contexts, individuals' preferences and evaluations are generally cognitively driven and based on the degree to which the product or service is a means to an end. On the contrary, hedonic consumption is based on experiential and sensory benefits and refers to products and services that bring pleasure and enjoyment through their use or possession. In these cases, consumers' preferences and responses are emotionally driven and based on the degree to which the service or product is rewarding. For example, gaining knowledge by reading a science magazine can be defined as a utilitarian form of consumption, while having fun by reading an entertainment magazine may reflect a hedonic type of consumption.

Longoni and Cian (2022) showed that people believe that AI technologies (recommendation systems or AI agents) are more competent than humans in assessing utilitarian attributes and information; therefore, they trust them more when they offer recommendations that refer to those types of services. The authors named this phenomenon 'word-of-machine effect' and showed that it is mainly based on the (erroneous) beliefs that AI and robots are rational and logical. In contrast, humans are associated with emotions and sensations. Interestingly, the effect holds in 'pure' utilitarian contexts and when utilitarian attributes are emphasized in a service or product. For example, companies can sell a chocolate cake (an inherently hedonic product) by highlighting its functional properties (e.g., chemical ingredients, healthiness, nutrients) or by highlighting its sensorial attributes (e.g., taste, aroma). In this case, a chocolate cake suggested by AI will be more successful when the AI stresses its utilitarian attributes of it. This can be relevant for food delivery services using AI to provide personalized customer offers. In this case, a personalized offer would be more successful if the offered product's utilitarian attributes were emphasized.

Second, AI-based services positively affect customers' responses and experience when they offer services that serve *objective* tasks (Castelo et al., 2019). Objective tasks are defined as tasks that involve measurable and quantifiable facts, such as analyzing data, scheduling events, or predicting the weather. Subjective tasks, on the contrary, are facts open to interpretation and based on opinion or intuition and include activities such as writing news articles or composing music. The main difference between these two tasks is the type of capabilities needed to perform them properly; the first (objective) would require more cognitive abilities, while the second (subjective) would need more affective and emotional abilities. However, as research on perceived humanness (e.g., Gray et al., 2007) has demonstrated, robots and generally AI technologies are seen as having some degree of cognitive abilities (i.e., the ability to plan and think) but lacking emotional and sensational abilities. Thus, individuals generally believe that AI provides better services when involved in objective tasks because they believe that AI is better equipped with the cognitive and rational abilities associated with these tasks (Castelo et al., 2019). Further, as in the previous case, the perceived objectivity of a task can be manipulated by increasing the degree to which a given task is perceived as objective. For example, a job candidate's evaluation could be done using standardized tests and measures or by conducting informal interviews. Thus, framing an AI-based service as an objective may increase its acceptance and usage, enhancing its customer experience.

The type of product or service the AI offers can also impact customers' responses. Granulo et al. (2021), for example, found that when a service has higher *symbolic value* and is strongly associated with *self-expression* motives, individuals prefer humans over AI technologies. For example, customers prefer a human-made poster of a body if it is used to decorate their places (i.e., expressing something about themselves) rather than illustrating the body's anatomy. This happens because customers use self-expressive products and services to express their uniqueness. While humans and human labor are associated with unique and varied products, AI-based services are generally perceived as reliable and consistent, thus more uniform. Similarly, customers are reluctant to adopt AI-based services when such services are strongly linked to their own identity (Leung et al., 2018). For example, if individuals consider themselves as 'proud bakers' and baking is an activity perceived as part of their identity, they will probably enjoy performing this activity themselves, resisting the automation of intelligent machines.

Finally, AI-based services can enhance the customer experience in service encounters where individuals can be exposed to *judgments* by the service providers (Pitardi et al., 2021; Holthöwer & van Doorn, 2022). Pitardi et al. (2021), for example, found that individuals feel less embarrassed and uncomfortable when they have to buy an embarrassing product (i.e., antifungal treatment for genital) from an AI agent than a human assistant. Similarly, Holthöwer and van Doorn (2022) showed that individuals are more likely to choose an AI agent than a human agent when they want to acquire an embarrassing product. As in the case of the task performed by the agent, this effect is driven by individuals' beliefs on the perceived capabilities of the technology. Specifically, robots and AI technologies are perceived as having a low degree of agency (i.e., ability to act with intentions, form opinions, and make moral and social judgments) and to lack emotions (i.e., ability to feel and sense; Gray et al., 2007). During embarrassing encounters, this perceived inability of AI agents to form opinions and make moral and social judgements on others' behaviors and purchases help relieve the discomfort, enhancing the overall customer experience. For example, fitness Apps such as Under Armour that utilize AI to offer personalized health programs benefit customers because individuals feel more at ease receiving diet and physical activity recommendations from a non-judgmental AI rather than from a fellow human.

These findings can be relevant for companies that use AI to personalize their services. For instance, AI personalized services such as the Sephora Visual Artist could be more effective and positively affect the overall customer experience when used for more objective and utilitarian products and services (i.e., grooming products) rather than for more hedonic ones (i.e., fragrances) Similarly, streaming platforms such as Netflix and YouTube may highlight AI-based recommendations when, for instance, viewers are more interested in utilitarian types of product (i.e., documentaries).

The Customer: Age, Gender, and Technology Readiness

Individuals characteristics are likely to change the overall experience of AI-based services. To date, literature has identified three relevant elements that can influence acceptance and usage of AI technologies: age, gender, and level of technology readiness.

When considering the *age* of the consumers, studies found that older customers in healthcare settings benefit from AI services thanks to the technology's ability to engage socially with users. For example, AI-based services for the elderly enhance customer experience by offering safeguarding, cognitive, and social support (Čaić et al., 2019). At the same time, older people are generally more averse to AI-based technology because of their digital inexperience and need for human contact

(Belanche et al., 2021). Children instead appear to benefit from AI-based services, especially when the technology is implemented in educational as well as rehabilitation and health care settings. For example, Sesame Street, the American educational children's television series, has recently incorporated IBM Watson AI to create an AI-vocabulary learning application. The app uses the students' vocabulary and reading level to recommend personalized learning exercises, and it updates its recommendations as the students progress in their learning. When looking at the *gender* of the customers, research shows that men tend to display more positive attitudes to AI technology than women (Chen & Huang, 2016). For example, fathers are keener than mothers in using educational AI-service for their children.

Individuals' propensity to embrace and use cutting-edge technology also impacts their experience. The literature identifies *Technology readiness* as a pivotal consumer characteristic that can either increase or demote the usage of technology-based services (Blut & Wang, 2020). Technology readiness is generally enabled by individuals' degrees of innate innovativeness and optimism. Customers who are more open to new things and hold a positive and favorable view of technology tend to be more likely to adopt and use new technologies. Discomfort and insecurity are detractors that lower individuals' degrees of technology readiness. Specifically, people who perceive a lack of control over the technology and are skeptical about the technology's ability to work properly tend to avoid the technology. For example, preliminary studies found that individuals with higher technology readiness perceive higher service quality when interacting with AI robots and are more likely to visit the service (Yoganathan et al., 2021). Given the AI's ability to offer interactive and social services, the degree of technology readiness becomes particularly relevant to understanding and improving the customer experience with AI-based services.

Macy's, an American department store, offers an interesting example of how consumer technology readiness and general personal inclinations can be considered when implementing AI-driven personalized services. The company uses the IBM Watson AI technology to power *Macy's On Call*, an AI digital assistant that guides customers through the store experience by providing recommendations and directions when they enter the store. The technology is trained to 'sense' whether the customers are frustrated or irritated with the AI service and, in case, immediately alert a human staff employee to support the customer. By doing so, Macy's can modulate its services based on the immediate individuals' reactions to the technology while respecting individual differences and avoiding enforcing the technology on not-ready customers.

The Technology: AI Design and Configuration

A final set of features that are likely to influence how individuals experience AI-based services are the design characteristics of the technology. Based on the configuration of such characteristics, AI applications can take various forms and differ in appearance, modality, and tangibility.

Most research on AI design focuses on understanding the role played by *anthropomorphism* and human-like features (Blut et al., 2021). To date, findings show that designing AI agents to look human can be a double edge sword. For example, some scholars found that when they are 'too much' human-like, robots can elicit consumer discomfort (Mende et al., 2019) and decrease consumer attitudes toward the technology and the service due to increasing feelings of eeriness and uncanniness (Kim et al., 2019). Also, due to their appearance, humanoid robots can generate a certain level of expectancy in terms of warmth and result in greater service dissatisfaction when such expectations are not met (Choi et al., 2021). On the other hand, studies showed how human-like features positively influence consumer responses via socio-cognitive evaluations (Belanche et al., 2021), resulting in higher trust and enjoyment (van Pinxteren et al., 2019). Humanoid robots, for instance, increase perceptions of service quality, visit intentions, and willingness to pay for a service (Yoganathan et al., 2021). Further, they are perceived as able to recover from common service failures by offering explanations and apologies (Choi et al., 2021).

More recently, studies in this area have taken a more sophisticated approach and analyzed how much AI agents look like humans and how much they act like humans influence the customer experience. These studies generally refer to the *mind attribution* theory (Gray et al., 2007) and examine how perceptions of the mind in AI agents—expressed in terms of the ability to think (agency) and ability to feel (experience)—impact customers' responses. So far, studies have found that AI agents are perceived as having a low level of agency and no experience (Söderlund, 2022). Low perceptions of mind in AI agents can have negative consequences for customers. For instance, individuals are more likely to accept brand misconduct and respond positively to brand harm crises caused by AI agents because the lower perceptions of mind lower the overall attributions of responsibility (Srinivasan & Sarial-Abi, 2021). However, low perceptions of the mind can also benefit the customer experience by facilitating service interactions in settings where individuals fear other people's judgments, such as during an embarrassing encounter (Pitardi et al., 2021; see section above).

AI agents can also have an assigned *gender*. For example, voice assistants are generally females (e.g., Amazon Alexa and Apple Siri), while financial advisors usually have male connotations (e.g., Cleo). Gender is one of the most important human characteristics influencing how people develop relationships. Scholars are trying to understand how such features in AI agents can affect customer responses. So far, research has shown that people apply social stereotypes (e.g., gender stereotypes) to AI agents and, as a result, tend to perceive female robots as warmer, friendlier, and more trustworthy than male robots (e.g., Borau et al., 2021; Stroessner & Benitez, 2019). Recently, Pitardi et al. (2022) found that not gender per se but rather the match between people's gender and the gender of the AI can enhance the overall customer experience in terms of positive attitudes and perceived comfort by increasing perceptions of control in the interaction with the technology. Thus, while there is an overuse of female gendering in robotics (e.g., Alexa, Siri, Sophia), adopting both male and female AI agents can increase positive responses toward the service and facilitate the overall customer experience.

Finally, AI agents can interact with customers through different modalities (i.e., text, voice, and haptic feedback) and with various degrees of *interactivity*. Voice-based AI agents generally elicit more social responses from users. Being perceived as inherently more human, they foster a greater sense of social presence that increases the levels of attraction, trustworthiness, and acceptance of the agent (Pitardi & Marriott, 2021).

Moreover, voice-based AI agents are generally more likely to be perceived as 'intelligent' and 'skillful' and promote a flow-like customer experience, resulting in more positive evaluations and behavioral outcomes (Zierau et al., 2022). However, when AI agents communicate through voices, their communications can also be perceived as more difficult to understand (Jeng et al., 2013). Moreover, since voice interactions are 'openly spoken,' they can increase privacy risks and prompt consumers' concerns (Melzner et al., 2020).

As for conversational *interactivity*, virtual chatbots that emulate natural human-to-human conversations lead to more positive evaluations of the service and more positive onboarding experiences (Hildebrand & Bergner, 2021). Thanks to the use of natural language that allows for real-time interactions, such conversational chatbots increase perceptions of affective trust resulting in greater acceptance of the AI-based service. Hence, developing conversational (virtual or embodied) AI-based service may increase its acceptance and usage, enhancing its customer experience.

The BMW intelligent personal assistant offers examples of various AI configurations and personalized services, an AI conversational assistant incorporated in the car that offers drivers personalized recommendations (i.e., about restaurants along the way or places to stop) while learning their preferences and habits.

Implications

The proposed framework identifies three main dimensions that may help marketers identify the conditions where AI-based services have the potential to positively influence the customer experience, in turn creating value for companies and businesses. While the dimensions have been presented one by one for explanatory purposes, they act simultaneously, interacting with each other. Their interplay allows us to identify the best fit between the context, the technology configuration, and customer characteristics. When adopting such an integrated perspective, important implications arise for various marketing decisions and society.

AI-Based Services Implementations

From the literature analyzed, it is clear that individuals' beliefs about whether a technology is effective in performing a certain activity are fundamental determinants of whether that technology is ultimately adopted. Being AI-based services associated with more cognitive and rational abilities, successful strategies to increase the customer experience will include implementing such technologies in objective and utilitarian types of services, where customers will trust the AI more and will perceive it as more competent. Moreover, AI-based services may be more successful in settings and encounters that customers perceive as uncomfortable or where unpleasant self-conscious emotions such as guilt, shame, and shyness can arise. For example, customers in a restaurant could feel less shy to send an unsatisfactory dish back when a robot is serving them. Similarly, in contexts such as funeral parlors, customers may benefit from neutral, low emotionally charged interactions offered by AI-based services. Further, when deciding what type of AI application should be used, conversational human-like agents of both genders are more likely to enhance the customer experience by fostering social responses and greater perceptions of control.

AI-Based Service Targeting and Positioning

When using AI-based services, factors such as customer technology readiness and age represent important information that needs to be considered in segmentation and targeting strategies. For example, AI agents developed for social and healthcare assistance can be more successful among older customers and children. Managers should also identify in their targets those groups of customers who are highly involved in specific product and service categories and, thus, more likely to perceive the same products and services as part of their identity (e.g., proud bakers). For such customers, the automation of the service through the adoption of AI technology can hurt their overall experience and relationship with the service provider.

Moreover, the technology readiness index can provide businesses with indications of whether the deployment of AI applications for specific groups of individuals can increase or reduce the customer experience. Finally, companies should prioritize a more functional positioning over an experiential one by emphasizing the utilitarian attributes of their service. For example, recommendation systems such as TripAdvisor could highlight the use of AI-based suggestions when the utilitarian characteristics are more relevant, as in business travel, and deemphasize it when suggesting leisure travel.

Implications for Individuals and Society

While an overview of the implications of AI-based service for individuals and society, in general, is out of the scope of this chapter, three important factors need to be mentioned. First, personalization comes with a price, and AI-based services constitute a potential threat to individuals' data privacy that can negatively impact the customer experience (Nazareno & Schiff, 2021). These technologies are by nature able to record, process and store information from their interactions, which results in higher data privacy and security concerns, especially when sensitive information is involved. Second, AI-based services can threaten consumer well-being by leading individuals to engage in low-quality decision-making, ultimately leading to dissatisfaction (Banker & Khetani, 2019). For instance, people are more likely to accept a poor product offering (i.e., a Uber ride twice as expensive as expected) from an AI agent because they perceive the offer as less exploitative due to lower perceptions of agency (Garvey et al., 2022). Third, AI-based services can incentivize unethical behavior because low perceptions of mind in the agent reduce feelings of guilt, thus encouraging individuals to cheat and lie (Kim et al., 2022). This is particularly relevant for companies operating in the service and public sectors, such as insurance companies and revenue agencies, where consumer dishonesty can damage the organization, consumers, and society. When looking at strategies to improve customer experience, such negative implications must be considered to minimize and avoid AI implementations that can potentially harm the final customers.

Conclusion

AI-based services are already part of our daily lives and will become even more pervasive shortly. Hence, companies and marketers must understand how adopting such technology can enhance the customer experience throughout the customer journey and lead to greater engagement and satisfaction. In this chapter, we identify three main dimensions—the context, the customer, and technology—whose interplay generates the conditions under which AI-based services may create or destroy business value. Marketers can develop appropriate strategies and techniques to deliver effective and engaging AI-based services by identifying the most fruitful fits between the technology's properties and configurations, the customers' needs and characteristics, and the contexts. However, the framework also allows us to recognize the conditions under which such technological interactions may destroy business value and constitute potential threats to customers' well-being and society in general. Relevant implications for individuals and society may arise from the 'blind' adoption of AI-based services, and companies and businesses must keep them in account when developing customer experience strategies.

References

Abdulquadri, A., Kieu, T. A., & Nguyen, N. P. (2021). Digital transformation in financial services provision: Perspective to the adoption of chatbot. *Journal of Enterprising Communities: People and Places in the Global Economy, 15*(2), 258–281.

Banker, S., & Khetani, S. (2019). Algorithm overdependence: How the use of algorithmic recommendation systems can increase risks to consumer well-being. *Journal of Public Policy & Marketing, 38*(4), 500–515.

Belanche, D., Casaló, L. V., Schepers, J., & Flavián, C. (2021). The humanness-value-loyalty model examines the effects of robots' physical appearance, warmth, and competence in frontline services. *Psychology & Marketing, 38*(12), 2357–2376.

Blut, M., & Wang, C. (2020). Technology readiness: A meta-analysis of conceptualisations of the construct and its impact on technology usage. *Journal of the Academy of Marketing Science, 48*(4), 649–669.

Blut, M., Wang, C., Wünderlich, N. V., & Brock, C. (2021). Understanding anthropomorphism in service provision: A meta-analysis of physical robots, chatbots, and other AI. *Journal of the Academy of Marketing Science, 49*, 1–27.

Borau, S., Otterbring, T., Laporte, S., & Fosso Wamba, S. (2021). The most human bot: Female gendering increases humanness perceptions of bots and acceptance of AI. *Psychology & Marketing, 38*(7), 1052–1068.

Čaić, M., Avelino, J., Mahr, D., Odekerken-Schröder, G., & Bernardino, A. (2019). Robotic versus human coaches for active ageing: An automated social presence perspective. *International Journal of Social Robotics, 12*(4), 867–882.

Castelo, N., Bos, M. W., & Lehmann, D. R. (2019). Task-dependent algorithm aversion. *Journal of Marketing Research, 56*(5), 809–825.

Chen, N. H., & Huang, S. C. T. (2016). Domestic technology adoption: Comparison of innovation adoption models and moderators. *Human Factors and Ergonomics in Manufacturing & Service Industries, 26*(2), 177–190.

Choi, S., Mattila, A. S., & Bolton, L. E. (2021). To err is human (-oid): How do consumers react to robot service failure and recovery? *Journal of Service Research, 24*(3), 354–371.

Daqar, M. A. A., & Smoudy, A. K. (2019). The role of artificial intelligence on enhancing customer experience. *International Review of Management and Marketing, 9*(4), 22.

De Keyser, A., & Kunz, W. H. (2022). Living and working with service robots: A TCCM analysis and considerations for future research. *Journal of Service Management., 33*(2), 165–196.

Dietvorst, B. J., Simmons, J. P., & Massey, C. (2015). Algorithm aversion: People erroneously avoid algorithms after seeing them err. *Journal of Experimental Psychology: General, 144*(1), 114.

Garvey, A. M., Kim, T., & Duhachek, A. (2022). Bad news? Send an AI. Good News? Send a Human. *Journal of Marketing*. https://doi.org/10.1177/00222429211066972

Granulo, A., Fuchs, C., & Puntoni, S. (2021). Preference for human (vs robotic) labour is stronger in symbolic consumption contexts. *Journal of Consumer Psychology, 31*(1), 72–80.

Gray, H. M., Gray, K., & Wegner, D. M. (2007). Dimensions of mind perception. *Science, 315*(5812), 619.

Hildebrand, C., & Bergner, A. (2021). Conversational robo advisors as surrogates of trust: Onboarding experience, firm perception, and consumer financial decision making. *Journal of the Academy of Marketing Science, 49*(4), 659–676.

Holthöwer, J., & van Doorn, J. (2022). Robots do not judge: Service robots can alleviate embarrassment in service encounters. *Journal of the Academy of Marketing Science*, 1–18.

Huang, M. H., & Rust, R. T. (2018). Artificial intelligence in service. *Journal of Service Research, 21*(2), 155–172.

Huang, M. H., & Rust, R. T. (2021). Engaged with a robot? The role of AI in service. *Journal of Service Research, 24*(1), 30–41.

Jeng, W., He, D., & Jiang, J. (2013). Users' perceived difficulties and corresponding reformulation strategies in voice search. In *The 7th annual symposium on human-computer interaction and information retrieval.* University of Pittsburgh.

Kaplan, A., & Haenlein, M. (2019). Siri, Siri, in my hand: Who's the fairest in the land? On the interpretations, illustrations, and implications of artificial intelligence. *Business Horizons, 62*(1), 15–25.

Kim, T. W., & Duhachek, A. (2020). Artificial intelligence and persuasion: A construal-level account. *Psychological Science, 31*(4), 363–380.

Kim, T. W., Lee, H., Kim, M. Y., Kim, S., & Duhachek, A. (2022). AI increases unethical consumer behaviour due to reduced anticipatory guilt. *Journal of the Academy of Marketing Science*, 1–17.

Kim, S. Y., Schmitt, B. H., & Thalmann, N. M. (2019). Eliza in the uncanny valley: Anthropomorphising consumer robots increases their perceived warmth but decreases liking. *Marketing Letters, 30*(1), 1–12.

Kumar, V., Rajan, B., Venkatesan, R., & Lecinski, J. (2019). Understanding the role of artificial intelligence in personalised engagement marketing. *California Management Review, 61*(4), 135–155.

Lee, M. K. (2018). Understanding perception of algorithmic decisions: Fairness, trust, and emotion in response to algorithmic management. *Big Data & Society, 5*(1), 2053951718756684.

Leung, E., Paolacci, G., & Puntoni, S. (2018). Man versus machine: Resisting automation in identity-based consumer behaviour. *Journal of Marketing Research, 55*(6), 818–831.

Liu-Thompkins, Y., Okazaki, S., & Li, H. (2022). Artificial empathy in marketing interactions: Bridging the human-AI gap in affective and social customer experience. *Journal of the Academy of Marketing Science, 50*, 1–21.

Logg, J. M., Minson, J. A., & Moore, D. A. (2019). Algorithm appreciation: People prefer algorithmic to human judgment. *Organizational Behavior and Human Decision Processes, 151*, 90–103.

Longoni, C., Bonezzi, A., & Morewedge, C. K. (2019). Resistance to medical artificial intelligence. *Journal of Consumer Research, 46*(4), 629–650.
Longoni, C., & Cian, L. (2022). Artificial intelligence in utilitarian vs. hedonic contexts: The "word-of-machine" effect. *Journal of Marketing, 86*(1), 91–108.
Melzner, J., Bonezzi, A., & Meyvis, T. (2020). *Verba Volant Scripta Manent: Communication modality affects privacy expectations.* ACR North American Advances.
Mende, M., Scott, M. L., van Doorn, J., Grewal, D., & Shanks, I. (2019). Service robots rising: How humanoid robots influence service experiences and elicit compensatory consumer responses. *Journal of Marketing Research, 56*(4), 535–556.
Mogaji, E., & Nguyen, P. N. (2022). Managers' understanding of artificial intelligence in relation to marketing financial services: Insights from a cross-country study. *International Journal of Bank Marketing, 40*(6), 1272–1298.
Mogaji, E., Soetan, T., & Kieu, T. (2020). The implications of artificial intelligence on the digital marketing of financial services to vulnerable customers. *Australasian Marketing Journal, 29*(3), 235–242.
Nazareno, L., & Schiff, D. S. (2021). The impact of automation and artificial intelligence on worker well-being. *Technology in Society, 67*, 101679.
Önkal, D., Goodwin, P., Thomson, M., Gönül, S., & Pollock, A. (2009). The relative influence of advice from human experts and statistical methods on forecast adjustments. *Journal of Behavioral Decision Making, 22*(4), 390–409.
Pitardi, V., Bartikowski, B., Osburg, V. S., & Yoganathan, V. (2022). Effects of gender congruity in human-robot service interactions: The moderating role of masculinity. *International Journal of Information Management, 70, 102489.* https://doi.org/10.1016/j.ijinfomgt.2022.102489
Pitardi, V., & Marriott, H. R. (2021). Alexa, she's not human, but… unveiling the drivers of consumers' trust in voice-based artificial intelligence. *Psychology & Marketing, 38*(4), 626–642.
Pitardi, V., Wirtz, J., Paluch, S., & Kunz, W. H. (2021). Service robots, agency and embarrassing service encounters. *Journal of Service Management, 33*(2), 389–414.
Russell, S. J., & Norvig, P. (2009). *Artificial intelligence: a modern approach.* Prentice Hall.
Söderlund, M. (2022). Service robots with (perceived) theory of mind: An examination of humans' reactions. *Journal of Retailing and Consumer Services, 67*, 102999.

Srinivasan, R., & Sarial-Abi, G. (2021). When algorithms fail: Consumers' responses to brand harm crises caused by algorithm errors. *Journal of Marketing, 85*(5), 74–91.

Stroessner, S. J., & Benitez, J. (2019). The social perception of humanoid and non-humanoid robots: Effects of gendered and machine-like features. *International Journal of Social Robotics, 11*(2), 305–315.

van Doorn, J., Mende, M., Noble, S. M., Hulland, J., Ostrom, A. L., Grewal, D., & Petersen, J. A. (2017). Domo arigato Mr. Roboto: Emergence of automated social presence in organisational frontlines and customers' service experiences. *Journal of Service Research, 20*(1), 43–58.

van Pinxteren, M. M. E., Wetzels, R. W. H., Rüger, J., Pluymaekers, M., & Wetzels, M. (2019). Trust in humanoid robots: Implications for services marketing. *Journal of Services Marketing, 33*(4), 507–518.

Wirtz, J., Patterson, P. G., Kunz, W. H., Gruber, T., Lu, V. N., Paluch, S., & Martins, A. (2018). Brave new world: Service robots in the frontline. *Journal of Service Management, 29*(5), 907–931.

Yeomans, M., Shah, A., Mullainathan, S., & Kleinberg, J. (2019). Making sense of recommendations. *Journal of Behavioral Decision Making, 32*(4), 403–414.

Yoganathan, V., Osburg, V. S., Kunz, W. H., & Toporowski, W. (2021). Check-in at the Robo-desk: Effects of automated social presence on social cognition and service implications. *Tourism Management, 85*, 104309.

Zierau, N., Hildebrand, C., Bergner, A., Busquet, F., Schmitt, A., & Marco Leimeister, J. (2022). Voice bots on the frontline: Voice-based interfaces enhance flow-like consumer experiences & boost service outcomes. *Journal of the Academy of Marketing Science*, 1–20.

6

Engaging Self-service in a Customer Service Ecosystem

George Spais

Introduction

Self-service technology (SST) research is a well-established study area with many studies. According to Google Scholar results, many works on SST have been published in the last ten years (7920 from 2010 to 2020, June 21, 2023). However, as of July 25, 2022, this research topic increased exponentially during the pandemic (2010 Google Scholar results for 2021–2022), following Wang et al. (2022). Their study found that customers are already familiar with the concept of SST and are affected by the pandemic differently. The behavioral implications of using SST during the pandemic were discussed (Wang et al., 2022). A literature search using the keywords "self-service technology" and "pandemics" returns 816 Google Scholar results as of July 25, 2022, confirming the observation. The "contactless economy" appears to be what distinguishes the COVID-19 crisis from earlier crises. Experts from all over the world

G. Spais (✉)
Hellenic Open University, Patras, Greece
e-mail: spais.georgios@ac.eap.gr

© The Author(s), under exclusive license to Springer Nature Switzerland AG 2023

J. N. Sheth et al. (eds.), *Artificial Intelligence in Customer Service*,
https://doi.org/10.1007/978-3-031-33898-4_6

concur that the new normal is emerging. For example, in the US, at-home consumption has the potential to be a $3 trillion market by 2025, according to Deloitte's 2022 report (Contactless Economy: Are You Prepared?).

Some trends are happening more quickly because of the crisis. Both supply-side factors (such as the development of digital technologies such as 5G, cloud platforms, artificial intelligence (AI), and data analytics) and demand-side factors are driving a contactless economy (e.g., the need for convenience, health, and safety). Customer service systems powered by AI are advancing because they are likely to improve productivity and customer experience at lower costs (Chen et al., 2021). Due to recent AI innovations, customers are eager to learn about digital marketing more before purchasing (Singh et al., 2019). Based on consumer's morality and social response theory, the crucial role of consumers' profiles based on the social reaction types (SRCM view); and the necessary guiding principles, we will develop a proposed model for planning the development of a self-service automation system in retailing to improve the AI experience with the deontological moral framework as the fundamental basis of the system based on a priori and a posteriori consumer knowledge before launching the enabling technologies (underlying the importance of consumer ethics and morality).

The "critical realistic view" of two critical examples is vital in this discussion. Amazon's cashier-less Go Grocery retail locations and the adoption of fully autonomous self-checkout systems ("Grab and Go technology") is the first critical example. The second critical example is the Computers Are Social Actors (CASA) paradigm with consumers on the Taobao app, the merchant's customer service chatbot agent (Song et al., 2022), and the perceived communication quality. Customers are conscious when interacting with an AI customer self-service system because they anticipate that robots will be just as capable of effective online communication as people. Customers' tech skepticism about predictive algorithms might cause substantial push-back by threatening their autonomy (e.g., Carmon et al., 2019). Thus, the initial assumption of the chapter is that consumers' adoption of technical frames—captured in perceptions about AI—is influenced by contextual reality and consumers' social responses to AI. Because end-user perceptions of technology may differ from those of the designers, it is helpful to look at how the incongruence may lead to conflicts in use, how consumers may react to such conflicts, and what results emerge from such interaction.

For marketers, selecting the wrong self-service options may harm customer loyalty. Automation of increasingly challenging tasks is doomed to failure. These problems require a more thorough inquiry and possibly several contacts, which an individual is far better equipped to handle. Due to the exciting research and potential theoretical, managerial, and social implications, integrating the two critical examples may bring high value to the topic of consumer engagement in customer self-service platforms driven by AI.

The chapter is organized into four sections. In the next section, "Background and Literature Review," we illustrate the broad definitions, operationalization, and discussions of the topic, and past literature, to demonstrate our position on the topic. The rest of the sections are "Background and Literature Review", "Self-service Automation, Customer Engagement, and Consumer AI Experience", "The Frame of Reference, Theoretical Framework, and Critical Examination of Current Examples", "Building and Engaging Self-service Automation System to Improve Customer Experience: Critical Examination of Current Examples", "The Proposed Model of Planning the Development of a Self-Service Automation System in Retailing to Improve the AI Experience Incorporating the SRCM View and Consumer Morality", "Implications". The last section "Conclusions" summarizes the conclusions of the chapter.

Background and Literature Review

In this section, we provide the necessary broad definitions with a short critical discussion to support the position on the topic.

The Benefits of AI-Powered Technologies

Artificial intelligence driven by technology is an innovation that transforms numerous organizational procedures and provides businesses and other entities with numerous advantages that greatly ease daily operations by utilizing machine learning and robots. A significant technical innovation that is revolutionizing numerous processes is AI-powered technology. Artificial intelligence is designed to change the current dynamics by providing substantial advantages to all facets of life to address the current inefficiencies in the performance of various functions (Mogaji & Nguyen, 2022). The most significant benefits that AI offers marketing departments (by greatly simplifying life using machine learning and robots) are:

Virtual assistants and two-way dialogic interactions with their customers: The emergence of virtual assistants (e.g., service robots or chatbot services) is one of the key benefits of artificial intelligence that has altered how many organizational functions (Abdulquadri et al., 2021). Now, customers can receive prompt, accurate responses from programmed virtual assistants. Another benefit is that virtual assistants typically offer individualized input (e.g., Jiang et al., 2022).

Minimizing analysis inaccuracy: Human error can occur in most jobs conducted by persons, meaning that a significant portion of the analysis may contain inaccuracies. Artificial intelligence (AI)-powered technology is more trustworthy than using humans to perform the analysis (e.g., Chung et al., 2020).

Predictive analytics: One of the key benefits of technology powered by AI is predictive analysis. This technique makes it possible for AI to perform forecasting, which is essential for making future predictions. Organizations can better foresee the future and make plans due to AI technology that provides predictive analysis (Mogaji et al., 2020). Because they lack the knowledge and tools to forecast the future or even guess what might happen to their organization, most businesses fail (e.g., Dubey et al., 2020).

Bulk processing: Every month, organizations handle substantial volumes of data to make decisions that will aid in running the business. The issue is that a massive volume of data needs to be processed, making it difficult to manage. Technology with artificial power facilitates bulk processing. This expression denotes the ability to quickly process vast amounts of data and assist the business in making informed management decisions. AI technology can handle large amounts of data much faster than a person ever could (e.g., Ilmudeen, 2022).

Cost-effectiveness: Many business owners believe that putting artificially powered technologies into use will cost their organization a lot of money. However, they overlook that this technology boosts business efficiency while reducing expenses in other operations. Although AI-powered technology may initially cost more, it can eventually help an organization save money (e.g., Huang & Rust, 2022).

Self-service Automation, Customer Engagement, and Consumer AI Experience

According to Lim et al. (2022), customer experience strategy will advance through a stronger emphasis on omnichannel and deeper integration of online and offline marketing channels. It indicates that self-service automation will become more critical over the coming years. Self-service automation is the automation of manual, time-consuming IT services to boost productivity and reduce manual procedures in providing IT services. According to Gartner, businesses can cut operational expenses by 30% by integrating hyper-automation technology into operating procedures by 2024.

Big data have lately driven the rapidness in innovations that support a sustainable transition to social robots, mobile internet, the internet of things, and AI-powered technologies (e.g., Kim et al., 2022). One of these automation models is self-service automation, which enables non-IT staff to use IT services without help from the IT team. Many marketers are looking for artificial intelligence (AI) and, logically, the necessary support from their IT teams to complete more tactical activities that do not require human experience. Chatbots managed by artificial intelligence are proving to be advantageous to organizations. Stores that utilize AI-powered marketing perform five times better than retailers in the traditional economy. Digital marketing has caused a significant shift in consumer behavior, with today's consumers wanting a more consistent and individualized experience. According to Salesforce, about 30% of marketing executives employed AI in 2018, but that percentage is expected to rise to 84% by 2020 (Davis, 2018). According to IDC (2022), by the end of 2026, global investment in AI-centric systems will pass $300 billion.

Based on the above, we anticipate that AI will fundamentally change service companies, affecting how businesses interact with their customers—and the key concept is "consumer AI experience" (CAIE). In recent years, customer engagement (CE), which assesses how consumers connect with brands, products, and businesses, has become one of many managers' critical strategic priorities. Research on AI and CE has mainly been undertaken in isolation, giving little insight into how the two domains interact, even though both fields have significant literature-based gains (Hollebeek et al., 2021). The three main elements of the artificial intelligence ecosystem are output systems (AI), gathering and storing of data, and statistical and computational techniques. Among the remarkable works in the AI marketing literature is the work of Puntoni et al. (2021), claiming that AI could be a neutral tool to be judged on its effectiveness and accuracy. However, this perspective ignores any potential societal and personal problems that might occur. It makes it possible for goods and services to carry out tasks generally thought to demand intellect and independent judgment on the part of consumers. Sociological and psychological scholarship must also be integrated to study some of the customers' expenses incurred in their interactions with AI. As a result, Puntoni et al. (2021) distinguished four categories of AI-related consumer experiences: data acquisition, categorization, delegation, and social. Consumers may be served or exploited by data collection experiences, understood or misunderstood by categorization experiences, empowered or replaced by delegation experiences, and connected or alienated by social experiences.

Clients are eager to spend more time educating themselves before purchasing due to AI innovation in digital marketing (Singh et al., 2019). Further, designing Marketing AI systems so that tech naysayers focus their emphasis mostly on privacy issues. However, robust prediction algorithms can face strong opposition since they endanger consumer autonomy (CA) (e.g., Carmon et al., 2019). Therefore, we can see that promising dynamic to grow into a potent service research substream in the years to come with the interplay and interconnections among AI, CE, CA, and CAIE in the self-service context.

The Frame of Reference, Theoretical Framework, and Critical Examination of Current Examples

The following frame of reference (built by the author) helps the readers deepen their understanding of the links and interrelationships of the clustered theories and critical theories and concepts in the AI literature in different sectors and realize the focus of this chapter (see boxes colored yellow). The frame of reference seeks an insightful and thorough explanation of AI literature's various theoretical and conceptual underpinnings. We outline the relevant clusters of theories and concepts in Fig. 6.1. Compared to other AI-powered technologies and self-service automation theories and models. The following frame of reference includes the social response theory (the adopted theory) and SRCM as the adopted strategic

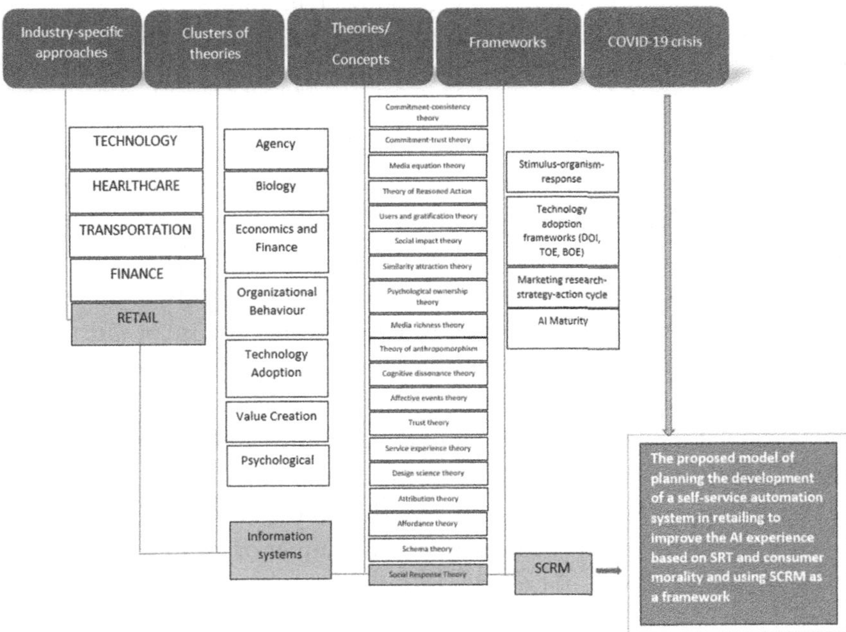

Fig. 6.1 The related theories, constructs and concepts in the frame of reference. (Source: Author)

framework. The adopted theory and the model (as the strategic framework) will allow proposing a model of continuous planning of the development of a self-service automation system in retailing to improve the AI experience incorporating consumer morality based on the experience that the critical examination of current examples (during the COVID-19 crisis).

Social Response Theory, Customer/Consumer Morality, and Critical Principles for Building an Engaging Self-service Automation System to Improve Customer/Consumer Experience

The most recent literature on AI in retailing underlines the value of social response theory regarding questions such as the replacement of human customer service by AI and the effect of poor communication and privacy concerns on the adoption of AI (e.g., Song et al., 2022).

The social response theory (introduced by Nass et al., 1994) explains how social interactions and human–technology interactions are comparable. According to the social response theory, when technology has a set of traits identical to people, an individual's response to the technology will appear as social behavior and be met with social regulations. According to the social response theory, if a user notices that a technological interface conveys positive social cues, this should encourage more interpersonal communication techniques and increase immersion. Previous studies have demonstrated that consumers' sense of satisfaction and involvement can increase when they fully engage in technology (Huang et al., 2020). Thus, as customers experience higher feelings of being sucked into anthropomorphized technology and wasting time, this should result in more positive opinions (Barney et al., 2022).

According to social response theory, after a service failure, AI responds to customers/consumers in an impersonal manner or even not at all (e.g., Lv et al., 2022). It will not only fail to eradicate the destructive emotions caused by the failure. The customer's negative perceptions and responses to AI services could be amplified, leading to uncontrollable, unpleasant emotions like anxiety and impatience. Understanding customer needs and seeing the situation from their perspective are only two examples of how the empathic reaction, a fundamental emotional component of the contact process, shows specific human emotional features and indicators. Customers are consequently more engaged and are more likely to respond favorably to AI automatically.

Another interesting extension of the social response theory and the AI-powered self-service automation systems that must receive the appropriate attention regarding the critical principles for building an engaging self-service automation system (that improves AI customer experience) is consumers' morality. It implies the moral evaluation of AI versus human agents, following Giroux et al. critical questions in morality (2022). Many works have now looked into how consumers and machines interact (e.g., Caruelle et al., 2022). First, people can believe that inorganic creatures, like robots and AI agents, have minds. This verdict may significantly impact how those actors' and people's impressions of their moral conduct are perceived (e.g., Bigman et al., 2019). The idea of mind is frequently linked to morality and upholding moral principles. Moral judgments frequently depend on the mental abilities of the agents (such as their ability to recognize the effects of their actions). In accordance with Giroux et al. (2022), consumers can believe that robots and other technologies can perceive minds. Additionally, a growing body of research has hypothesized computers as social actors, suggesting how consumers respond to technology from a social psychological perspective (e.g., Cheng et al., 2022).

Social Response Context Model (SRCM)

One of the prevailing models of social response theory is the Social Response Context Model (e.g., Mahardhika et al., 2016), as a cognition model analyzing mutual beliefs and mental subgroupings for explication and communication purposes. The model must consider an individual's public and private positions before and after an effect to account for the spectrum of responses to social influence. (Nail & MacDonald, 2011; Nail et al., 2000).

The following four questions can be used to acquire the SRCM's dimensions of AI-powered self-service automation systems following Nail et al. (2000). Before being exposed to influence, what are consumers' public position toward a potential influence source (preexposure, public agreement/disagreement)? What was the consumers' private opinion before being influenced (private agreement/disagreement, preexposure)? After being exposed to influence, what are the consumers' public reactions toward the influence source (postexposure, public agreement/disagreement)? What are the consumers' private reactions (postexposure, private agreement/disagreement) after being influenced? Combinatorial analysis with all four questions results in 16 fundamental responses that can be influenced. The authors adopted the four-letter descriptors for explication and communication purposes. The letters represent the agreement (A) or disagreement (D) on each pertinent dimension. Before influence, the first letter stands for the public dimension, while the second stands for the intimate dimension. After influence, the third and fourth letters represent the public and private dimensions.

There are six social reaction types that the model includes (1) "congruence" (AA/AA), (2) "disinhibitory contagion" (DA/AA), and (3) the four responses outlined in Allen's (1965) model ["conversion" (DD/AA), "compliance" (DD/AD), "paradoxical anticompliance" (DD/DA), and "independence" (DD/DD)]. Along with "paradoxical anticompliance" (DD/DA), the model also includes "anticompliance" (AA/DA) and "anticonversion" (AA/DD; see Nail et al., 2000, p. 457) as two other forms of "anticonformity."

Building an Engaging Self-service Automation System to Improve Customer Experience: Critical Examination of Current Examples

We regard the necessary breakdown as essential to expand our understanding of and explain the investigated conditions (see Fig. 6.2) and based on the eight fundamental principles of critical realism, in accordance with Sayer (1992) ("object/entities: businesses and organizations,"

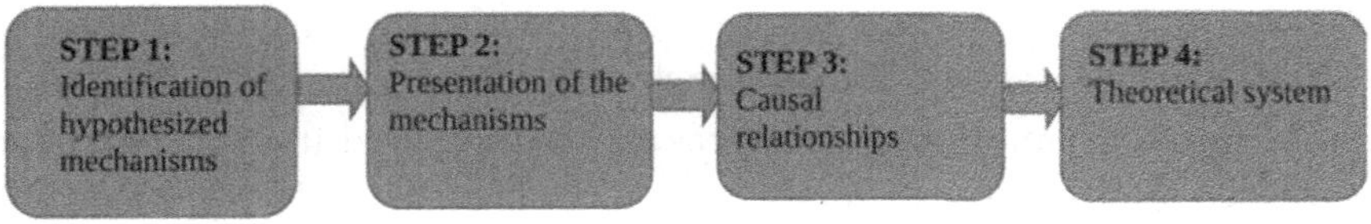

Fig. 6.2 Steps of our critical realism approach based on Spais and Paul (2021), p. 328

"causal power: SRCM strategic approach," "context: COVID-19 crisis" and "events: successful performance").

Amazon GO Concept in the Pandemic (The USA and Canada): The largest online retailer in the world, Amazon, is poised to revolutionize shopping, even offline. The pandemic helped Amazon enter the brick-and-mortar market, and its cashier-less technology is expected to become the next big thing in retail. The critical example reports how individuals feel and what they anticipate from tech-enabled retail, like Amazon Go. Amazon developed the cashier-less concept, the first company to design, test, and provide this kind of store in the retail sector; a store concept was introduced in December 2016. Customers can pick up what they need and leave the store without going through the checkout procedure thanks to the concept, which uses computer vision and machine learning.

Amazon Go Concept in the Pandemic: Consumers' Public Toward the Potential Influence Source and Consumers' Private Opinion Before Being Influenced

According to a survey (December 2016), although the idea is intriguing, American consumers were not convinced that Amazon Go would make their lives easier. Amazon's technology has gained most American adults' trust and is eager to try it (Richter, 2018). Many people, however, are reluctant to pay more for a checkout-free shopping experience because they think it would exacerbate rather than resolve problems.

Five years of testing and implementation were required to conclude. The year 2018 saw the debut of Seattle's first Amazon Go store and the associated Just Walk Away App. Intending to build 3000 new stores by the end of 2021, they subsequently launched 25 additional outlets in the US. Two years after the initial launch, Amazon decided to expand and build Amazon Go Grocery, a larger store that allowed users to purchase packaged goods and fresh foods (e.g., Gazzola et al., 2022). The interactions with technology included interaction with the Amazon App and machine learning, computer vision, sensor fusion, and "just walk out technology" (regarding AI). An automatized service, where customers only interact with the app to enter and exit the store. The idea was the introduction of a technology focused on convenience, and the service was intended to have no contact with employees. The only chance of human interaction is the case of technology malfunction or misunderstanding. Customer experience is frictionless thanks to no lines and checkout cashiers, with no interaction and enhancement through the encounter. Some adverse reactions included 1) no real-time shopping list update and not "thank you for shopping with us!". The entire experience is a deliberate devaluing of the benefits of human interaction.

With plans to open hundreds of Amazon Go outlets around the country in the coming years, the retail colossus will not only challenge Walmart and Kroger's supremacy in the physical arena. However, it will deliver the "Just Walk Out" shopping experience to the customers' front doors.

Based on the SRCM typology of reactions, how avidly are Americans and Canadians anticipating the idea of no queues and no checkout before and during the pandemic?

Amazon Go Concept: Consumers' Public Reaction Toward the Influence Source After Being Exposed to the Influence

Piplsay conducted a nationwide survey of 30,166 to get the information. According to a Pipslay survey of US citizens 18 years of age and older performed nationwide in 2021, 57% of people will be thrilled to see an Amazon Go or another tech-enabled store nearby. About 60% of Americans, 65% of Millennials, and 60% of Gen Xers believe Amazon Go would put leading brick-and-mortar shops like Walmart and Kroger in danger. Despite the entrenchment of internet retail, half of Americans think that stores like Amazon Go will succeed. Comparatively, 35% of Millennials and Gen Xers and 30% of Gen Zers believe that Amazon Go-style stores will be a success. Fifty-four percent of men will be more thrilled about nearby tech-enabled stores like Amazon Go than women.

Data from Dalhousie University less than a year ago (2021) showed that, for the first time, Canadian shoppers were beginning to favor self-checkouts when exiting the grocery store, as reported by Charlebois (2022). Self-checkouts are expected to be used frequently by fewer than 53% of Canadians in the future. The projections show that millennials and Gen Z (born between 1997 and 2005 and 1981 and 1996, respectively) use these self-service tills more frequently than younger consumers. According to CivicScience, only 19% of customers older than 55 felt ready to use self-checkouts before the pandemic, compared to 35% of people in the "35 to 54" age group. For all demographics at the time, cashiers were still the chosen choice ("anticonversion": AA/DD). These machines used to be loved or hated by consumers during the current crisis, including a percentage of anti-conformist consumers (not recorded in the survey). Serviced checkouts are losing ground to self-checkouts as their popularity soars. A startling 75% of Canadians used self-checkouts at least once in the previous six months, according to a survey done by Dalhousie University in collaboration with Caddle in early May.

Furthermore, 85% of Canadians indicated they were pleased with their experience ("congruence": AA/AA). Notably, 47% of Canadians said they would go to a grocery shop without cashiers. All purchases are recorded by digital sensors, allowing customers to add what they want to their basket and exit the store without ever using a checkout. The most well-known example of this at the moment is Amazon Go. The amount was far smaller before the pandemic. Consumer acceptance of technology at the grocery store is growing, including not only the consumers that show "congruence" but also the percentage of consumers that show "disinhibitory contagion" (DA/AA), as grocers dealing with the uncomfortable optics of witnessing machines taking employment. Cashiers were heroes during the pandemic, and everyone wanted to give them a raise (Charlebois, 2022). Further, the political representatives have even condemned the big chains for discontinuing compensation plans to provide workers with improved working circumstances. However, the reality is that it is still challenging to find and keep good employees. With the current labor scarcity, it is even worse. The agri-food business is gradually prioritizing automation and robotics, particularly in the food service and retail sectors. Once choices have been made, the majority of customers do not want to stand in line to make payments.

In Figs. 6.3 and 6.4, we present two tree diagrams to apply SRCM for Amazon Go for Canadians. Figure 6.3 shows the application of the model for the majority of Canadians categorized in "congruence" and "anticonversion" after being exposed to the influence. Figure 6.4 shows the application of the model for the Canadians categorized in "disinhibitory contagion" after being exposed to the influence.

CASA Paradigm with Consumers on the Taobao App and Merchant's Customer Service Chatbot Agent

Taobao is an online marketplace in China, introduced by the Alibaba Group introduced Taobao, a C2C and B2C e-commerce platform, in 2003. It provides individuals and small-business owners a platform to market their goods online to customers mostly in Chinese-speaking countries and internationally through their own online stores. Its retailers typically offer express delivery services. When selling their items, sellers can choose a predetermined price or an auction. Most transactions are set

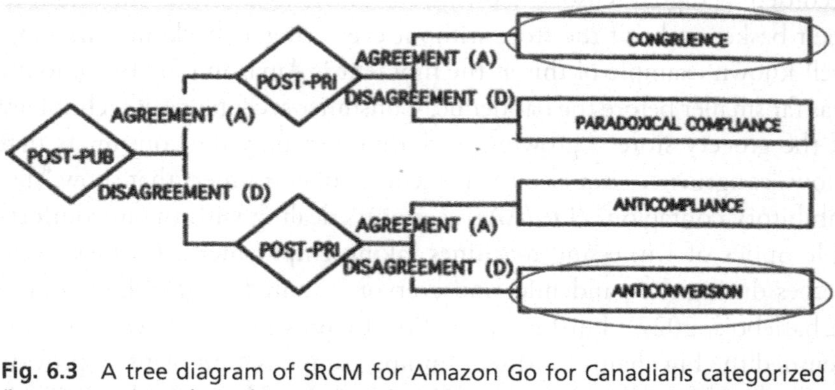

Fig. 6.3 A tree diagram of SRCM for Amazon Go for Canadians categorized in "congruence" and "anticonversion" based on MacDonald et al. (2004), p. 80

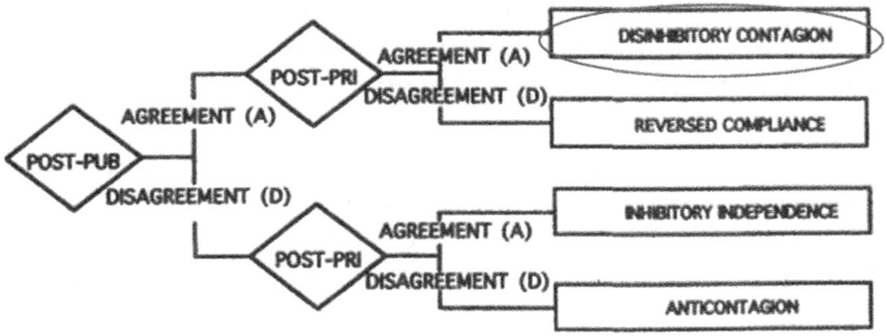

Fig. 6.4 A tree diagram of SRCM for Amazon Go for Canadians categorized in "disinhibitory contagion" based on MacDonald et al. (2004), p. 80

prices, with auctions accounting for a small share. Buyers can review seller profiles on the internet using ratings, comments, and complaints. Taobao's widely used payment system is Alipay from Alibaba (Huang et al., 2021). The vendor lists a good or service in their Taobao store on the website. Vendors can list their products and services on Taobao and sell them at fixed rates or in auctions. Most offerings are different goods with fixed prices (similar to traditional retail stores).

Customers can look up product and seller information, reviews, and ratings to evaluate the goods and retailers. The customer searches for particular products using the Taobao search engine, and customers will message the shop using Taobao's instant messaging service if he has any queries (Ali Wang Wang). The customers use Alipay, Alibaba's online and mobile payment system, and may use Escrow to keep this money (for up to ten days). The buyer must get the items from the vendor after the seller receives the payment notification. After getting the things, the customer can assess their suitability and give the product a rating. The funds will then be released from the Escrow and given to the merchant. Taobao hosts several festivals of online purchasing per year (the most well-known being Single's Day or Double Eleven) to promote stores and products. In addition to paid search, live streaming, and other marketing options, these festivals give vendors a chance to advertise their goods and draw in Chinese customers.

CASA: Consumers' public position Toward the Potential Influence Source and Consumers' Private Opinion Before Being Influenced

The influence of service agent type (chatbots vs. humans) on consumer adoption intention was examined by Song et al. (2022). This experiment divided the service agent type's altered material into two components. Based on textual material, the questions inviting the participants to imagine their wish to buy a piece of home décor through an online platform, such as "what would be the interface between you and the retailer's customer service?". The second section consisted of visual stimuli that show a customer interacting with the merchant's customer service team using the Taobao app, human, and chatbot agents. The Wenjuanxing platform was used to find people for an online survey that would help determine the experimental stimuli in more detail. The platform is used to find willing research study volunteers online and is comparable to the Amazon Mechanical Turk site. Approximately 70% of participants were between the ages of 18 and 40. In the pre-experiment, clothing items were divided into five clothing categories. For the most frequent forms of clothes, subjects were asked to respond to two questions measuring their familiarity with the products. To disprove that gender affects the experiment's results, "home clothing" was selected as the experimental stimulus. The findings showed that gender, professional, and performance clothing did not influence casual, sports, and residential attire.

CASA: Consumers' Public Reaction Toward the Influence Source After Being Exposed to the Influence

In their first experiment, Song et al. (2022) showed that the service agent type significantly impacted consumers' adoption intentions. Consumers were more open to scenarios involving human interactions than chatbot interactions. In the second experiment, Song et al. (2022) examined variations in consumers' willingness to accept various service agent types based on perceived communication quality characteristics. There is a view that the COVID-19 pandemic has significantly altered individual and organizational communication practices, resulting in significant social change. This difference is evident while interacting in the virtual world. In total agreement with recent works (e.g., Bag et al., 2021), AI has a positive relationship with user engagement and conversion. The experiments' results show that conversion satisfies AI experience ("conversion" DD/AA).

CASA: Consumers' Private Reaction After Being Influenced

Comparing the effectiveness of Twitter communication between a bot and a human agent, Song et al. (2022) discovered that there is no distinction between a bot and a human agent in terms of source credibility, interaction intentions, or communication proficiency using the CASA paradigm, by Edwards et al. (2014). In terms of communication and interactional intentions, the bot is also seen as appealing, credible, and competent. Lastly, the human agent was seen as more alluring than the bot in social and task contexts (Edwards et al., 2014). Suppose users' personality and communicative features differ when they first engage in human–human interaction and human-–artificial intelligence (i.e., chatbot) interactions. Users confirmed that they exhibit different personality and communicative traits when interacting with human–human and human-–artificial intelligence. They are especially cautious while interacting with people because they are "more open, more conscientious, more extroverted, more agreeable, and self-disclosing than AI" (Mou & Xu, 2017).

A dialogue occurred between the customer, the merchant's chatbot, and human customer service representatives on the Taobao app. The image depicted the human beings in the scenario asking the customer service three questions and receiving a humanized response. When the subject likewise asked customer service these three questions in the chatbot scenario, the picture suggested that the subject got official responses.

Song et al. (2022) found that the service agent type significantly impacted consumers' adoption intentions in the first experiment. Consumers were more open to scenarios involving human interactions

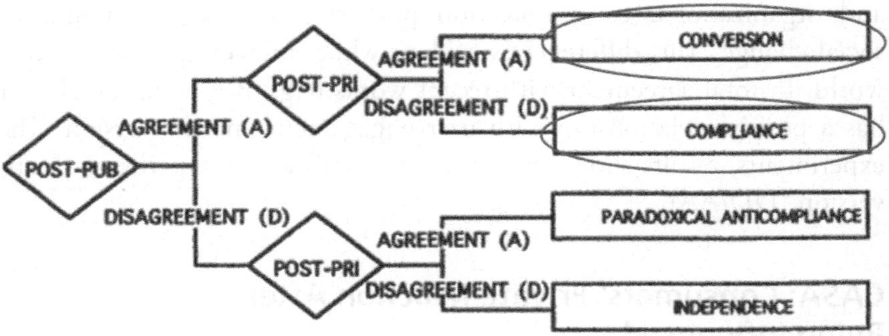

Fig. 6.5 A tree diagram of SRCM for Amazon Go for CASA paradigm with consumers on Taobao app and merchant's customer service chatbot agent based on MacDonald et al. (2004), p. 80

than chatbot interactions. The authors addressed these issues in their second experiment. Based on the five elements of perceived communication quality, they looked at variances in consumers' willingness to accept different service agent kinds. Additionally, the results show that social presence mediates the influence of anthropomorphic design signals on consumers' "compliance" (DD/AD), following recent works (e.g., Adam et al., 2021).

In Fig. 6.5, we present a tree diagram to apply SRCM for the CASA paradigm with consumers on the Taobao app and merchant's customer service chatbot agent categorized in "conversion" and "compliance" after being exposed to the influence.

The Proposed Model of Planning the Development of a Self-Service Automation System in Retailing to Improve the AI Experience Incorporating the SRCM View and Consumer Morality

A new era has arrived in which humans are starting to build artificial intelligence (AI) machines, integrating human morals and ethics, carrying out jobs, and making judgments based on the moral principles we teach them. It is taking place as the existing literature looks at how morality affects ethical judgments. Kant created deontology, which emphasizes the underlying ethical standard for making judgments rather than the results of such conclusions. Previous studies on digital personal assistants have also focused on gender issues (Søndergaard & Hansen, 2018), politics influenced by economic aspects (Ojeda, 2021), and regulating bots on speech rights (Lamo & Calo, 2019; Bendel, 2019). Furthermore, the above studies do not investigate the morality and ethics underlying the knowledge these chatbots provide to customers. The moral frameworks that AI programmers utilize may vary depending on the situation (Murray, 2020).

Based on the two critical examples analyzed above, in Fig. 6.6, we present our six-step proposed model of continuous planning of the development of a self-service automation system in retailing to improve the AI experience of customers/consumers using SRCM as a framework and under the prism of consumer morality and the ten guiding principles for improving self-service automation system in a retailing context. The steps are as follows.

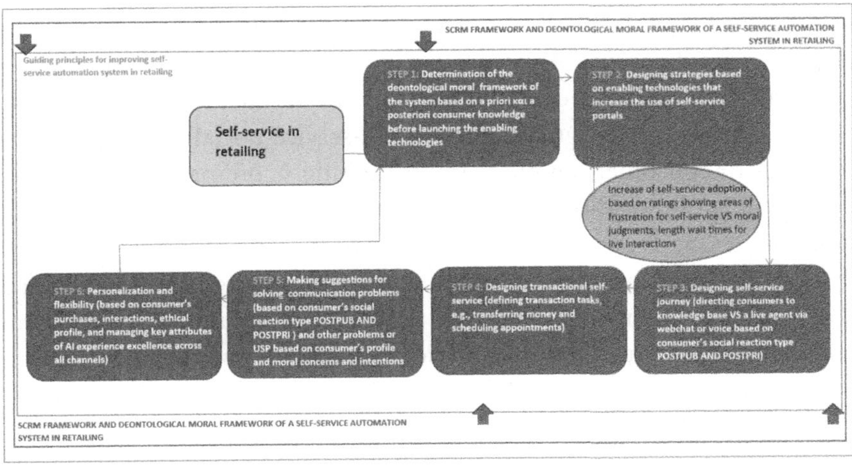

Fig. 6.6 The proposed model of continuous planning of the development of a self-service automation system in retailing to improve the AI experience incorporating the SRCM view, consumer morality and the guiding principles (Source: Author)

Step 1: Determination of the Deontological Moral Framework of the System

A deontological perspective of ethics (rule-based) is compulsory in AI programming. Instead of the more calculative and character-based methods of consequentialism and virtue ethics. Businesses must meet the demand for a "computationally explicit trackable means of decision making." Businesses need deontology because interpretations that rely on evaluating marketing outcomes or making moral judgments are much more arbitrary and challenging to follow systematically. Humans must be able to transparently identify the reasoning behind a decision to appropriately assess the morality of the action in question since AI can make moral decisions that are exceedingly complicated.

Step 2: Designing Strategies Based on Enabling Technologies That Increase the Use of Self-service Portals

Managers frequently discover that developing and managing efficient self-service portals is more challenging than it looks as businesses hurry to deploy technology that allows customers to seek help independently. An excellent strategy for a company to start the installation of an IT self-service portal is to design an efficient self-service portal in the style of a business-to-consumer supplier. Implementing a pre-packaged solution with little customization could result in time, money, and resources on something that does not meet the businesses' or end customers' demands. It is up to companies to determine what influences customers' satisfaction and discontent with self-service portals. It also examines the problem of customer acceptance of self-service portals and emphasizes prerequisites for a client's initial use of a self-service portal.

Step 3: Designing a Self-service Journey

A chatbot can question customers about their needs and connect them to the self-service knowledge base to find the answers.

Step 4: Designing Transactional Self-service

When self-service is transactional, it becomes more complicated but also very beneficial. Since each of these tasks required the involvement of a human, they all helped lower the businesses' operational costs.

Step 5: Suggestions for Solving Communication and Other Problems or USP Based on the Consumer's Profile, Moral Concerns, and Intentions

It appears best to avoid discussing moral issues in conversations and steer clear of merely subjective and emotionally charged topics. The possibility of ethical problems occurring is typically associated with particular issues. Consumers who misinterpret moral and ethical concerns are a common problem for businesses. Morals reflect one's specific views on what is good or bad. However, the term "ethics" refers broadly to moral principles. However, it may refer to concerns about right and appropriate action in a much more specific context. AI could advise ways to solve a problem or sell something in line with the preferences or behaviors of the consumer.

Step 6: Personalization and Flexibility

Messaging is home to numerous channels with many comparable but slight differences. One significant hurdle to clear is being available on the media that the consumers/customer prefer. Still, it can be challenging to make this process simple for both them and the agents. Further, businesses should incorporate personalization and customization as they raise the bar for self-service because they are essential. Instead of only offering customers searchable information bases, offer some personalization based on that customer's prior interactions or purchases.

The ten guiding principles are based on a review of several recently published empirical works (e.g., Binder et al., 2022; Chen et al., 2022; Wang & Han, 2022):

1. Easily finding self-service for consumers and potential consumers. Potential consumers will get more frustrated by this lack of accessibility, leading to increased use of aided support channels, including phone, email, and direct contact with support professionals.
2. Giving consumers control and making self-service simple to use. The self-service businesses with the highest success rates can utilize self-service to empower, reassure, and build user confidence. Self-usability services are essential to its success. The success of the self-service portal application will be boosted by greater user adoption, improved user loyalty, less attrition, and word-of-mouth marketing when giving consumers the power to make decisions. In the self-service application, overwhelming customers with much information would only make them feel powerless.
3. Recognition of consumers' needs and ethical concerns. The ultimate goals of most self-service users are to solve an issue or obtain knowledge. An organization's knowledge of regularly arising issues may significantly influence self-service resolution rates. Retailers aware of these problems can design their self-service interactions so that consumers can address them and accomplish their objectives.
4. Easily readable content. By enriching the material with images, videos, and bolded language, consumer engagement will rise because doing so may force users to scan the page and look at these aspects. It is a great mistake to undervalue the significance of content because engaging material is essential to a positive user experience. By adhering to these content principles, the content will be more successful and more straightforward to translate if the retailers cater to a multilingual customer base.
5. Options for the omnichannel help desk. Not only is it crucial for the AI experience but it can also affect the effectiveness of self-service in retail by providing a consistent experience across numerous channels tailored to each customer.

6. Analysis and optimizations of consumer Feedback. Consider comments to be the judges or mentors who assist the business in achieving the ultimate objective of a fantastic self-service application using many available instruments and methods.
7. Self-service performance measurement. It is common practice to ignore this element of the customer's experience. All the graphics must be web-optimized to ensure the self-service application runs smoothly. It is crucial to gauge the business performance of the AI experience of self-service and its technical performance. This metric does not determine task success. A self-service session is deemed successful if consumers view any answers and no incidents are submitted for issues that might have been resolved by reading the FAQs or knowledge base articles rather than using support services.
8. Personalized experience. A customer's current condition and mental state can be revealed to the marketing team if they consider facts that they already know or that can be automatically collected.
9. Making accessibility support mandatory. Large clusters of consumers will be able to complete self-service tasks thanks to good accessibility, which ensures that the self-service information is distributed as widely as possible in the market. A technically inaccessible self-service site will provide low self-service rates to people with disabilities. All users benefit from good visual design, straightforward icon usage, and readable font sizes, but the elderly and others with vision impairments do.
10. Modification of consumer behavior. It is probably due to their mental model if some consumers' who need low-complexity, high-volume interactions choose a channel like a phone, which is quite expensive for the business. For instance, if a consumer has a mental model that suggests calling will be the easiest option, it may be because they have previously had negative experiences with self-service. The marketing teams can rebuild this mental model to reflect the actual path of least resistance by taking the time to "educate" issue users to use self-service by introducing them to the self-service experience. Techniques for behavior modification can also be utilized to eliminate rewards that may result from a customer's tendency to abuse the phone channel.

Implications

For the first time in the retail self-service literature, we propose a six-step model of continuous planning for the development of a self-service automation system in retailing to improve the AI experience (continuously) with the deontological moral framework as the fundamental basis of the system based on a priori and a posteriori consumer knowledge before launching the enabling technologies (underlying the importance of consumer ethics and morality). Further, the model incorporates the crucial role of consumers' profiles based on the social reaction types (SRCM view); and the necessary guiding principles. The proposed six-step model will allow retailers to build a self-service automation system that seeks to improve the AI experience (as the number one strategic priority in their marketing strategies). Based on the two critical examples, the retailers can achieve it by considering the most critical barriers and challenges to AI adoption and consumer engagement in self-services in retailing.

In the next two subsections, some interesting critical implications follow for industry, society, and academia.

Implications to Industry

Self-service machine operators can change their businesses into digitally based, intelligent environments that employ fewer people. Retailers are attempting to connect physical storefronts with online shopping using AI-powered solutions that can address integration problems by fusing digital and physical retail (e.g., Dwivedi et al., 2022). By enabling companies to build computer vision systems that can watch over Point Of Sales (POS) areas and automatically match items being scanned to corresponding Stock Keeping Units (SKUs), preventing fraudulent scans and Universal Product Code (UPC) barcode switching, artificial intelligence in the retail market helps prevent product loss. Customers can thus find what they need via self-service.

Algorithms can dynamically estimate demand for specific products or product categories by monitoring enormous volumes of data from business websites and online stores, ERP systems, social networks, and other sources. AI is used in physical and online retailers to improve supply chain operations and ensure that the most popular items are constantly in stock, even when demand is peaking (e.g., Kembro & Norrman, 2022).

Retailers desire more control over product supplies at every stage while remaining adaptable and agile in the face of difficulties as logistic chains get longer and more entangled, especially in light of the COVID-19 pandemic that has had a lasting effect on global business. Machine learning algorithms leverage a customer's purchase history, personal information, search history, and relevant datasets from the same user cohort to tailor product offers. The algorithms can also create personalized landing pages, which give customers the best, "hand-picked," depending on their previous purchases or the projected frequency of product renewal or replenishment, products they are most likely to acquire. These methods can be utilized by retail businesses developing their first online marketplace, modernizing an existing one, or striving to create a mobile app with advanced personalization.

Continuing the thoughts in the pandemic context, when considering how customers use SST in a pandemic environment, the selfish component of social relationships is raised because of social isolation's harmful effects. It implies that companies must confirm or deny the impact of the numerous hypothesized causes of consumers' use of SST. Businesses may recognize various emergent SST user segments and pandemic-related consequences on the exterior consumer environment and the interior self-perceptions of individuals, both of which impact how people use SST.

Implications to Society and Academia

Our daily lives have undergone a significant transformation due to artificial intelligence (AI), both personally and professionally. AI has assisted us in tracking the global spread of COVID-19 and forecasting what it will look like when it spreads further. AI is assisting us in determining what we want, sometimes even before we are aware of it (e.g., Lewis & Marsh, 2022), with everything from movie, cuisine, and music recommendations to suggested purchases on eCommerce websites. However, even more relevant and startling than the fact that we have grown accustomed to sharing our reality with intelligent simulations is the possibility that a technological exercise in exhibitionism could profoundly affect preserving civil liberties and human rights.

It is self-explanatory that a vast research area needs to be further explored through the lens of marketing. It will allow us to present a complete picture of the body of knowledge about artificial intelligence (AI) published in various bodies of literature (such as marketing, consumer research, psychology, and sociology literature). It will encourage interdisciplinary research teams and collaborations between business schools and schools of social sciences. Additionally, following recently published thoughts on a wheel of marketing education (e.g., Crittenden, 2022), business schools must update their marketing curriculums, especially for courses and modules analyzing AI-powered self-services and self-technologies, such as Retailing and Services Marketing. Regarding the societal role of business schools, they must develop new, readily available educational programs on retail artificial intelligence (AI) for anyone seeking a critical and unique perspective on how AI is infused into various aspects of our life, particularly as customers. Business schools must discuss pedagogical approaches to developing and assessing retail AI under an interdisciplinary view. Following are some queries regarding training citizens as consumers in the new AI retail context: Why share such consumer evolution? What is the punitive knowledge to be shared? What are the competencies to develop? How can it be discussed and evaluated?

Conclusion

The theoretical basis of the chapter is consumer ethics, morality, and social response theory, incorporating the crucial function of consumer profiles based on social reaction types (SRCM perspective) and the essential guiding principles. The development of fully autonomous self-checkout machines and Amazon's cashier-less Go Grocery retail outlets serve as the first critical example ("Grab and Go technology"). The second critical example is the CASA paradigm with consumers on the Taobao app, the merchant's customer service chatbot agent, and the perceived communication quality. The theoretical basis and the critical examples allowed us to propose a six-step model for planning the development of a self-service automation system in retailing.

In discussing the implications to retailers, it seems that retailers are seeking to combine digital and physical retail to link physical shops with online buying by utilizing AI-powered technologies to solve integration issues. While remaining flexible and nimble in the face of challenges as logistic chains get longer and more complicated, especially in light of the COVID-19 pandemic, retailers seek greater control over product supplies at every level.

Regarding the implications to academia, and especially to business schools, it seems that they must also rebuild their marketing curricula in light of recently published ideas on a "wheel of marketing education." Particularly for courses and modules that analyze AI-powered self-services and self-technologies like Retailing and Services Marketing. It will encourage interdisciplinary research teams and collaborations between business schools and schools of social sciences. Regarding the societal function of business schools, they must create new, easily accessible educational programs on retail AI for anyone looking for a critical and distinctive perspective on how AI is incorporated into various facets of our lives, particularly customers. Business schools must have interdisciplinary discussions about pedagogical methods for creating and evaluating retail AI. The discussion of the implications ended with critical questions about preparing regular people to act as customers in the emerging new AI-powered retail environment.

References

Abdulquadri, A., Kieu, T. A., & Nguyen, N. P. (2021). Digital transformation in financial services provision: Perspective to the adoption of chatbot. *Journal of Enterprising Communities: People and Places in the Global Economy, 15*(2), 258–281.

Adam, M., Wessel, M., & Benlian, A. (2021). AI-based chatbots in customer service and their effects on user compliance. *Electronic Markets, 31*(2), 427–445. https://doi.org/10.1007/s12525-020-00414-7

Allen, V. L. (1965). Situational factors in conformity. In *Advances in experimental social psychology* (Vol. 2, pp. 133–175). Academic Press. https://doi.org/10.1016/S0065-2601(08)60105-7

Bag, S., Srivastava, G., Al Bashir, M. M., Kumari, S., Giannakis, M., & Chowdhury, A. H. (2021). Journey of customers in this digital era: Understanding the role of artificial intelligence technologies in user engagement and conversion. *Benchmarking: An International Journal., 29*, 2074. https://doi.org/10.1108/BIJ-07-2021-0415

Barney, C., Hancock, T., Jones, C. L. E., Kazandjian, B., & Collier, J. E. (2022). Ideally human-ish: How anthropomorphized do you have to be in shopper-facing retail technology? *Journal of Retailing, 98*, 685. https://doi.org/10.1016/j.jretai.2022.04.001

Bendel, O. (2019). Chatbots as moral and immoral machines. In Implementing Artefacts in Machine Ethics. CHI 2019 Workshop on conversational agents, Glasgow, UK. https://www.robophilosophy.com/wp-content/uploads/2019/08/Paper_Chatbots_CHI.pdf.

Bigman, Y. E., Waytz, A., Alterovitz, R., & Gray, K. (2019). Holding robots responsible: The elements of machine morality. *Trends in Cognitive Sciences, 23*(5), 365–368. https://doi.org/10.1016/j.tics.2019.02.008

Binder, F., Diels, J., Balling, J., Albrecht, O., Sachunsky, R., Philipp, J. N., Scheurer, Y., Münsch, M., Otto, M., Niekler, A., Heyer, G., & Thorun, C. (2022). Putting users in the loop: How user research can guide AI development for a consumer-oriented self-service portal. In *International conference on human-computer interaction* (pp. 3–19). Springer. https://doi.org/10.1007/978-3-031-05434-1_1

Carmon, Z., Schrift, R., Wertenbroch, K., & Yang, H. (2019). Designing AI systems that customers won't hate. *MIT Sloan Management Review, 61*(2), 1–6.

Caruelle, D., Shams, P., Gustafsson, A., & Lervik-Olsen, L. (2022). Affective computing in marketing: Practical implications and research opportunities afforded by emotionally intelligent machines. *Marketing Letters, 33*(1), 163–169. https://doi.org/10.1007/s11002-021-09609-0

Charlebois, S. (2022, May 25). Majority of Canadians now embrace using self-checkouts at grocery stores: Study. Retail Insider. https://retail-insider.com/retail-insider/2022/05/majority-of-canadians-now-embrace-using-self-checkouts-at-grocery-stores-study/#:~:text=According%20to%20a%20survey%20conducted,were%20satisfied%20with%20their%20experience.

Chen, H., Chen, S., & Zhao, J. (2022). Integrated design of financial self-service terminal based on artificial intelligence voice interaction. *Frontiers in Psychology, 13*, 850092. https://doi.org/10.3389/fpsyg.2022.850092

Chen, J. S., Tran-Thien-Y, L., & Florence, D. (2021). Usability and responsiveness of artificial intelligence chatbot on online customer experience in e-retailing. *International Journal of Retail & Distribution Management, 49*(11), 1512–1531. https://doi.org/10.1108/IJRDM-08-2020-0312

Cheng, X., Zhang, X., Cohen, J., & Mou, J. (2022). Human vs. AI: Understanding the impact of anthropomorphism on consumer response to chatbots from the perspective of trust and relationship norms. *Information Processing & Management, 59*(3), 102940. https://doi.org/10.1016/j.ipm.2022.102940

Chung, M., Ko, E., Joung, H., & Kim, S. J. (2020). Chatbot e-service and customer satisfaction regarding luxury brands. *Journal of Business Research, 117*, 587–595. https://doi.org/10.1016/j.jbusres.2018.10.004

Crittenden, V. L. (2022). Thoughts on a wheel of marketing education. *Journal of Marketing Education, 44*(1), 3–5. https://doi.org/10.1177/02734753211070076

Davis, J. S. (2018). Introducing the 5th salesforce state of marketing report: Here are the top trends redefining the profession, Salesforce (December 13, 2018). https://www.salesforce.com/blog/introducing-fifth-state-of-marketing-report-blog/

Dubey, R., Gunasekaran, A., Childe, S. J., Bryde, D. J., Giannakis, M., Foropon, C., Roubaud, D., & Hazen, B. T. (2020). Big data analytics and artificial intelligence pathway to operational performance under the effects of entrepreneurial orientation and environmental dynamism: A study of manufacturing organisations. *International Journal of Production Economics, 226*, 107599. https://doi.org/10.1016/j.ijpe.2019.107599

Dwivedi, Y. K., Hughes, L., Baabdullah, A. M., Ribeiro-Navarrete, S., Giannakis, M., Al-Debei, M. M., et al. (2022). Metaverse beyond the hype: Multidisciplinary perspectives on emerging challenges, opportunities, and agenda for research, practice and policy. *International Journal of Information Management, 66*, 102542. https://doi.org/10.1016/j.ijinfomgt.2022.102542

Edwards, C., Edwards, P., Spence, A., & Shelton, A. (2014). Is that a bot running the social media feed? Testing the differences in perceptions of communication quality for a human agent and a bot agent on twitter. *Computers in Human Behavior, 33*, 372–376. https://doi.org/10.1016/j.chb.2013.08.013

Gazzola, P., Grechi, D., Martinelli, I., & Pezzetti, R. (2022). The innovation of the Cashierless store: A preliminary analysis in Italy. *Sustainability, 14*(4), 2034. https://doi.org/10.3390/su14042034

Giroux, M., Kim, J., Lee, J. C., & Park, J. (2022). Artificial intelligence and declined guilt: Retailing morality comparison between human and AI. *Journal of Business Ethics, 178*, 1027–1041. https://doi.org/10.1007/s10551-022-05056-7

Hollebeek, L. D., Sprott, D. E., & Brady, M. K. (2021). Rise of the machines? Customer engagement in automated service interactions. *Journal of Service Research, 24*(1), 3–8. https://doi.org/10.1177/1094670520975110

Huang, M. H., & Rust, R. T. (2022). A framework for collaborative artificial intelligence in marketing. *Journal of Retailing, 98*(2), 209–223. https://doi.org/10.1016/j.jretai.2021.03.001

Huang, F., Wong, V. C., & Wan, E. W. (2020). The influence of product anthropomorphism on comparative judgment. *Journal of Consumer Research, 46*(5), 936–955. https://doi.org/10.1093/jcr/ucz028

Huang, L., Zhang, X., Wang, Y., & Xie, X. (2021, September). Research and application of reinforcement learning recommendation method for Taobao. In *2021 IEEE Symposium on Computers and Communications (ISCC)* (pp. 1–6). IEEE. https://doi.org/10.1109/ISCC53001.2021.9631429

IDC. (2022). Worldwide spending on AI-centric systems will pass $300 Billion by 2026, according to IDC (September 22, 2022). https://www.idc.com/getdoc.jsp?containerId=prUS49670322.

Ilmudeen, A. (2022). Artificial intelligence, big data analytics and big data processing for IoT-based sensing data. In *Transforming management with AI, big-data, and IoT* (pp. 247–259). Springer. https://doi.org/10.1007/978-3-030-86749-2_14

Jiang, H., Cheng, Y., Yang, J., & Gao, S. (2022). AI-powered chatbot communication with customers: Dialogic interactions, satisfaction, engagement, and customer behavior. *Computers in Human Behavior, 134*, 107329. https://doi.org/10.1016/j.chb.2022.107329

Kembro, J., & Norrman, A. (2022). The transformation from manual to smart warehousing: An exploratory study with Swedish retailers. *The International Journal of Logistics Management, 33*(5), 107–135. https://doi.org/10.1108/IJLM-11-2021-0525

Kim, T., Jo, H., Yhee, Y., & Koo, C. (2022). Robots, artificial intelligence, and service automation (RAISA) in hospitality: Sentiment analysis of YouTube streaming data. *Electronic Markets, 32*(1), 259–275. https://doi.org/10.1007/s12525-021-00514-y

Lamo, M., & Calo, R. (2019). Regulating bot speech. *UCLA Law Review, 66*, 988. https://doi.org/10.2139/ssrn.3214572

Lewis, P. R., & Marsh, S. (2022). What is it like to trust a rock? A functionalist perspective on trust and trustworthiness in artificial intelligence. *Cognitive Systems Research, 72*, 33–49. https://doi.org/10.1016/j.cogsys.2021.11.001

Lim, X. J., Cheah, J. H., Dwivedi, Y. K., & Richard, J. E. (2022). Does retail type matter? Consumer responses to channel integration in omni-channel retailing. *Journal of Retailing and Consumer Services, 67*, 102992. https://doi.org/10.1016/j.jretconser.2022.102992

Lv, X., Yang, Y., Qin, D., Cao, X., & Xu, H. (2022). Artificial intelligence service recovery: The role of empathic response in hospitality customers' continuous usage intention. *Computers in Human Behavior, 126*, 106993. https://doi.org/10.1016/j.chb.2021.106993

MacDonald, G., Nail, P. R., & Levy, D. A. (2004). Expanding the scope of the social response context model. *Basic and Applied Social Psychology, 26*(1), 77–92. https://doi.org/10.1207/s15324834basp2601_7

Mahardhika, D., Kanno, T., & Furuta, K. (2016). Team cognition model based on mutual beliefs and mental subgrouping. *Journal of Interaction Science, 4*(1), 1–10. https://doi.org/10.1186/s40166-016-0014-6

Mogaji, E., & Nguyen, P. N. (2022). Managers' understanding of artificial intelligence in relation to marketing financial services: Insights from a cross-country study. *International Journal of Bank Marketing, 40*(6), 1272–1298.

Mogaji, E., Soetan, T., & Kieu, T. (2020). The implications of artificial intelligence on the digital marketing of financial services to vulnerable customers. *Australasian Marketing Journal, 29*(3), 235–242.

Mou, Y., & Xu, K. (2017). The media inequality: Comparing the initial human-human and human-AI social interactions. *Computers in Human Behavior, 72*, 432–440. https://doi.org/10.1016/j.chb.2017.02.067

Murray, R. (2020). The perceived ethics of artificial intelligence. *Markets, Globalization & Development Review, 5*(2), 3. https://doi.org/10.23860/MGDR-2020-05-02-03. https://digitalcommons.uri.edu/mgdr/vol5/iss2/3

Nail, P., & MacDonald, G. (2011). On the development of the social response context model. In *The science of social influence* (pp. 193–221). Psychology Press.

Nail, P., MacDonald, G., & Levy, D. A. (2000). Proposal of a four-dimensional model of social response. *Psychological Bulletin, 126*(3), 454. https://doi.org/10.1037/0033-2909.126.3.454

Nass, C., Steuer, J., & Tauber, E. R. (1994). Computers are social actors. *Human Factors in Computing Systems, 94*, 72–78. https://doi.org/10.1145/259963.260288

Ojeda, C. (2021). The political responses of virtual assistants. *Social Science Computer Review, 39*(5), 884–902. https://doi.org/10.1177/0894439319886844

Puntoni, S., Reczek, R. W., Giesler, M., & Botti, S. (2021). Consumers and artificial intelligence: An experiential perspective. *Journal of Marketing, 85*(1), 131–151. https://doi.org/10.1177/0022242920953847

Richter, F. (2018, January 22). What consumers think about Amazon Go. Statista. https://www.statista.com/chart/7198/consumer-opinion-on-amazon-go/.

Singh, J., Flaherty, K., Sohi, R. S., Deeter-Schmelz, D., Habel, J., Le Meunier-FitzHugh, K., & Onyemah, V. (2019). Sales profession and professionals in the age of digitization and artificial intelligence technologies: Concepts, priorities, and questions. *Journal of Personal Selling & Sales Management, 39*(1), 2–22. https://doi.org/10.1080/08853134.2018.1557525

Søndergaard, M., & Hansen, L. (2018). Intimate futures: Staying with the trouble of digital personal assistants through design fiction. In Proceedings of the *2018 Designing Interactive Systems Conference,* pp. 869–880. https://doi.org/10.1145/3196709.3196766.

Song, M., Xing, X., Duan, Y., Cohen, J., & Mou, J. (2022). Will artificial intelligence replace human customer service? The impact of communication quality and privacy risks on adoption intention. *Journal of Retailing and Consumer Services, 66*, 102900. https://doi.org/10.1016/j.jretconser.2021.102900

Spais, G., & Paul, P. (2021). A crisis management model for marketing education: Reflections on marketing education system's transformation in view of the Covid-19 crisis. *Marketing Education Review, 31*(4), 322–339. https://doi.org/10.1080/10528008.2021.1951120

Wang, S. M., & Han, C. (2022). An analysis of gender differences in the innovative function design of supermarket self-service checkout kiosk. In *International conference on human-computer interaction* (pp. 337–349). Springer. https://doi.org/10.1007/978-3-031-05544-7_26

Wang, X., Wong, Y. D., Sun, S., & Yuen, K. F. (2022). An investigation of self-service technology usage during the COVID-19 pandemic: The changing perceptions of 'self' and technologies. *Technology in Society, 70*, 102032. https://doi.org/10.1016/j.techsoc.2022.102032

7

Understanding the Role of Data in Artificial Intelligence-Based Personalization

Shubhadeep Mukherjee

Introduction

Artificial intelligence (AI) can be defined as "a system's ability to interpret historical data correctly, to learn from such data, and to use those learnings to achieve specific tasks and goals through flexible adaptations" (Haenlein & Kaplan, 2019). AI is affecting people's lives and businesses in unprecedented ways. AI in marketing has come to the forefront, with businesses providing personalization to consumers to meet specific consumer demands (Khrais, 2020). Thus, AI can potentially provide a competitive advantage to businesses. A big part of AI's application depends on how organizations can use data to solve complex customer problems. The arrival of big data capabilities has fundamentally transformed how organizations function, making data ubiquitous and inimitable for decision-making. The fundamental challenge remains on how the

S. Mukherjee (✉)
Operations and Information Management, Aston University, Birmingham, UK
e-mail: s.mukherjee3@aston.ac.uk

© The Author(s), under exclusive license to Springer Nature Switzerland AG 2023
J. N. Sheth et al. (eds.), *Artificial Intelligence in Customer Service*,
https://doi.org/10.1007/978-3-031-33898-4_7

plethora of data could be utilized to generate insights for decision-making. AI can use algorithms to learn from historical data and create customer intuitions leading to the mechanization of business practices (Davenport et al., 2020; Mogaji et al., 2020). Machine learning (ML) is a technology that has been the vanguard of data analysis for AI applications. Unlike other automation processes, which depend on the repetition of tasks, ML is a subset of AI that learns from past instances through historical data (Verma et al., 2021). ML can make decisions for an unseen problem based on the quality and quantity of those data points. For instance, based on the historical profile of people a bank has loaned to, ML can identify the riskier profiles of potential new customers. Another example could be the identification of objects on the shelf of a store. ML can predict whether the product belongs to the organization or a rival company based on the object's shape and color. This is possible because exposing the algorithm to many pictures makes it identify shapes and patterns. Along with ML, deep learning (DL) and natural language processing (NLP) technologies are used to analyze data, understand customer needs, and generate marketing insights (Davenport et al., 2020). Deep learning can be considered a subset of machine learning; however, some scholars disagree with this and consider deep learning a separate type of learning. Deep Learning (DL) entails the application of artificial neural networks with multiple layers to extract features from data. At the same time, NLP is a subset of AI that provides a computer with the ability to understand human language. Modern businesses are likely to use a combination of ML, DL, and NLP for data analysis and generating customer insights. Before one can analyze the data, the data must be collected and cleaned. Once the business problem has been defined and the data sources are identified, data collection and pre-processing (cleaning) become the most crucial steps in the data science life cycle. Once the data has been cleaned, we apply ML-based algorithms to draw actionable insights. This process is repeated for different business problems.

In this chapter, we develop a conceptual framework to explain the role of data in AI-based personalization. We discuss the aspects of data collection, data augmentation, data quality, data pre-processing, and ML/DL technologies application to solve business problems relating to personalization. We further address customer experience issues using data analytics, data and sustainability, responsible data science, data transparency, and the ethics of personalization through AI. Finally, we present a comprehensive and robust framework through an integrative literature analysis. We also provide implications for businesses and academicians along with the study's limitations.

Review of Literature

Recording browsing behavior, buying behavior, comments, and ratings can capture customer behavior on websites. Customer data may be structured or unstructured in nature. Based on the data's nature, it can be stored in a regular database system or data lakes. Further, based on the availability and the nature of the data, it could be enhanced by either discovery, augmentation, or generation; this entails the data collection part (Mogaji et al., 2020). The data collected is not in an ideal form to apply machine learning yet. Hence, it needs to be pre-processed and standardized. This is where we deal with outliers, missing values, normality, conversion from unstructured to structured data, etc. Once the data has been pre-processed, we apply ML/DL-based algorithms (recommender systems, sentiment analysis, etc.) to analyze it. Based on the quality and consistency of the results obtained, organizations can make their decisions and apply customer-specific personalization. This framework is similar to the Crisp-DM cycle's flow (Wirth & Hipp, 2000). We delve further into the framework in the results section discussing each construct in detail. We also discuss the components of data pre-processing in the results section (Fig. 7.1).

Structured and Unstructured Data

The arrival of big data technologies has also led to an increase in the diversity of data. Big data has been famously defined using the 5Vs (Sagiroglu & Sinanc, 2013). They are velocity (high speed of data generation), volume (Large data size), value, variety (diversity in data types), and veracity (accuracy or truthfulness of data). Evaluating data has

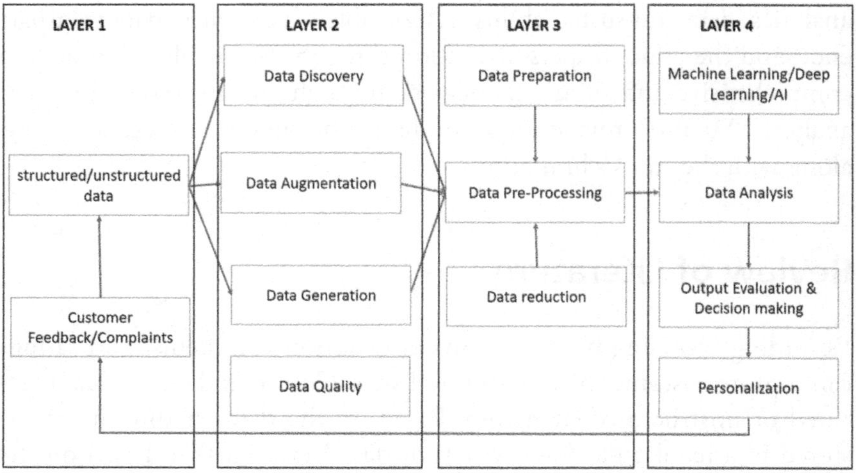

Fig. 7.1 Conceptual Framework for the Role of Data in AI for Personalization

become more complex due to the 5Vs; data variety has made the analysis more challenging. Data has been divided into three types, namely: structured, unstructured, and semi-structured. Structured data is the traditional type prevalent in the pre-big data era. It consists of numeric data, which is specific, ordered, and can be stored in traditional databases—for example, the tax return values of individual customers linked to their social security numbers. Unstructured data is the most dominant data form in the big data era. It consists of data that does not have a defined form or structure and is stored in its native form. Unstructured data cannot be stored in traditional databases and require specific storage designed for them. Examples of unstructured data could be text-based customer complaints on a website, sound bites from a music album, images on Google, etc. Analyzing unstructured data is a much more challenging task as compared to the analysis of structured data. Semi-structured data have some form or structure to them. For instance, emails, CSV files, and so on. Semi-structured data are not fit for analysis in their native form. However, they can be stored in some traditional databases.

Data Collection Using AI

Data collection entails the process of acquiring data to train models. Data collection for AI consists of three major ways: identification of the right data (data discovery), supplementing discovered data (data augmentation), and creation of data when none is available (data generation) (Roh et al., 2019). Here, we discuss the three methods one by one.

Data Discovery

For analysis to begin, data needs to be discovered. Hence, the researcher must share or search for new datasets on the web or in organizational data lakes (Halevy et al., 2016; Terrizzano et al., 2015). The critical tasks here are search scaling and identifying data appropriate for training. Collaborative analysis plays a critical role here. Different datasets can be shared, combined, and analyzed using a DataHub (Bhardwaj et al., 2015). Besides the data hub, which can be useful for collaborating within an organization, data can be shared over the web using fusion tables. Data journalism uses fusion tables extensively (Roh et al., 2019). One of the prominent web-based data-sharing platforms is Kaggle. Kaggle hosts a repository of thousands of datasets and even holds competitions where peers can upload their codes. Winners are decided based on the final performance of the machine learning algorithm. University of California Irvine (UCI) machine learning repository consists of more than 622 datasets that researchers and students can use for free. The datasets are classified based on the type of data science problems, data types, instances, and year of upload. Other than these, researchers can use various paid data platforms in return for a fee.

Data Augmentation

Data augmentation can enrich existing datasets by adding pre-trained or existing data to the new dataset. For instance, hidden or latent characteristics of data could be added to the datasets. One way latent features could be explored is Latent Dirichlet Allocation (LDA) (Blei et al., 2003). Using LDA, one can identify the hidden topics underlying a text document (Kumar et al., 2022), also known as topic modeling. Other techniques to augment data are Word2vec (Mikolov et al., 2013) which utilizes Continuous Bag-of-Words (CBOW), and skip-gram (Nguyen & Grishman, 2013). CBOW can predict the next word in a text based on the adjacent words. On the other hand, Skip-gram can predict the adjacent words based on the word in context. Word2vec represents a corpus of real words (Goldberg & Levy, 2014). Next, we discuss data generation.

Data Generation

When historical data is not available for a particular problem, data generation techniques could be applied to create data for training purposes. This can be done in two ways. One is to collect data through surveys or crowdsourcing (Daniel et al., 2018; Wang et al., 2016). Surveys have been an important source of data collection for empirical research (Couper, 2017). However, the data size required to train an ML algorithm sufficiently cannot be collected through surveys themselves. Hence, crowdsourcing of data becomes relevant. One of the most important sources of crowdsourced data is social media. Researchers have extensively used social media websites such as Twitter and Facebook to collect data for analysis (Mukherjee & Bala, 2017; Mukherjee & Bala, 2016; Kumar et al., 2022). These social media websites provide APIs to developers, which they can use to download data on various topics using hashtags. The data can also be downloaded for specific dates or between certain dates based on historical events. Another way used by authors to generate data was through Amazon Mechanical Turk, where tasks were assigned to human workers who were compensated on completion of the assigned work. Crowdsourcing has been the primary source of data labeling to run supervised learning tasks. In some cases, when data points are missing in a dataset, semi-supervised learning techniques are adopted to create data based on the trends in the dataset. Sometimes, measures of central tendency (mean, median, and mode) are also used to fill in missing quantitative data.

Data Analysis Using AI

Several AI-based techniques have been used in literature for analyzing data for personalization. We will go through the major techniques in this subsection, starting with machine learning.

Machine Learning

Personalization in customer service has become a necessity rather than a luxury (Goldenberg et al., 2021). Machine learning allows researchers and practitioners to access big data to create relevant, precise, and tailored customer recommendations. Based on the historical buying behavior of customers, machine learning-based algorithms, specifically recommender systems, can provide users with new recommendations (Mogaji & Nguyen, 2022; Verhagen et al., 2014). The personalization field has been dominated by collaborative filtering-based recommender systems largely based on matrix factorization. Compared to other areas of application of machine learning which perform exceedingly well, personalization problems are more complex to formulate and implement (Bernardi et al., 2019).

Deep Learning

Deep Learning (DL) entails the application of artificial neural networks with multiple layers to extract features from data. Deep learning-aided personalization in recommender systems' use is becoming prominent in academia (LeCun et al., 2015). There have been applications of deep learning-based systems for content-based, collaborative filtering, and hybrid models (content and collaborative). The major applications are in the field of entertainment, e-commerce, and news (Singhal et al., 2017). Content-based filtering has been used to predict song preferences using audio-based features of sounds using deep learning belief networks and probabilistic graphical models (Van den Oord et al., 2013; Wang & Wang, 2014).

Natural Language Processing

Natural Language Processing (NLP) is a subset of artificial intelligence that allows computers to understand human language through text and speech. NLP has applications in machine translation (from one language to another), global positioning system (GPS), and chatbots for customer service. Machine translation allows organizations to reach global outreach to brands in seconds, something unimaginable even a decade back (Abdulquadri et al., 2021; Seljan & Dunđer, 2014). NLP has played an unprecedented role in improving organizations' online presence through chatbots, question-answering systems, and real-time translation systems (Farzindar & Inkpen, 2015; Ramaswamy & DeClerck, 2018).

AI for Personalization

In the previous sections, we discussed applications of AI for data collection and analysis. In this section, we focus on personalization using AI. Personalization has been the core goal of marketing experts and practitioners for a long period. Personalization refers to having different or customized policies toward customers based on their past interaction with the brand or product. For example, the interaction could be based on earlier purchases, social media comments, browsing history, and so forth. An organization's ability to personalize its products for customers depends on its ability to differentiate between customers based on their traits (Rafieian & Yoganarasimhan, 2022). Big data plays a crucial role in personalization as larger data leads to better recommendations. This requires analysis of complex data patterns to identify what works better for the customers. Hence, we discuss the various AI-based methods through which personalization could work. We start with a basic introduction type of data and machine learning which are the basis of most AI-based learning.

Machine Learning for Personalization

IBM defines machine learning as "a branch of AI and computer science which focuses on using data and algorithms to imitate how humans learn, gradually improving its accuracy." An ML-based system's central goal is to learn from its mistake without human intervention. There are multiple ways to train historical data and identify patterns using ML. Prominent among them are supervised learning, unsupervised learning, and semi-supervised learning. Supervised learning is used when the analyst has information about the dependent variable (technically known as labeled data) (Caruana & Niculescu-Mizil, 2006). For instance, to predict the correct product segment for a customer based on historical data of other customers where the segment (high income, medium income, and low income) is already known, and the features are similar. Another example could be identifying a marketing mix based on customer feedback on social media. New customer feedback can be classified if historical information about the marketing mix component is available. Unsupervised learning information about the dependent variable is unavailable, and the analysis is primarily in the data exploration phase (Donmez et al., 2010). An example of this is identifying topics in customer complaints on social media. The complaints could be related to service quality, product quality, durability, and so forth. These can be automatically identified using topic modeling.

Another example could be the creation of word clouds based on the frequencies of words in customer complaints. The higher the frequency, the larger the size of the word. As the name suggests, semi-supervised learning can be considered an amalgamation of supervised and unsupervised learning techniques (Zhu, 2005). Here, some of the data for the dependent variable is labeled. However, most of it is unlabeled. The labeled data section within the dataset makes predictions for the unlabeled section. Another important technique within machine learning is sentiment analysis or opinion mining, which is used to identify customer sentiments (happy, sad, or neutral) about a service. In the next subsection, we discuss the major ML algorithms which can be utilized for personalization.

Recommender Systems

Recommender systems are a machine learning-based technique that aids computers in providing recommendations to users based on user similarity (with other users) or item similarity (with other bought items) (Ricci et al., 2011). Data plays an inevitably central role in recommender systems. Multiple rows of scalable data are required to understand and excavate similarity patterns among many users. These recommendations can be in the daily use items, online buying, and movie or music recommendations, among others. The major recommendation techniques are content-based (Lops et al., 2011; Thorat et al., 2015), collaborative filtering (Ekstrand et al., 2011; Schafer et al., 2007), and association rule mining (Kim & Yum, 2011).

Content-based filtering is grounded on the user's interaction with the products. Recommendations for users, in this case, are grounded on the accumulated knowledge of the user based on their browsing history and behavior (Van Meteren & Van Someren, 2000). For example, if a user rents action movies on YouTube, the YouTube algorithm will recommend similar action movies to the user. The filtering of recommendations is based on the similarity between the bought item and the recommended item. Although content-based recommender systems have huge applications in marketing, they require a massive amount of domain knowledge about the concerned person. The recommendations can go repetitive or boring as the method is based on feature similarity, which may give rise to a lack of novelty in recommendations.

The collaborative filtering technique is used to make recommendations based on similarity in users, which is grounded upon ratings or preferences of similar users (Schafer et al., 2007)—building upon the YouTube movie recommendation. For example, if three users have rented the same two movies on YouTube and a fourth user has rented one of the two. The second movie may be recommended to the fourth user based on similarity. The key idea is that users who have had similar interests in the past will continue to have so in the future. Collaborative filtering can be of two types: User-user-based filtering and item-item-based filtering. User-user-based filtering is when similar items are recommended to a user based on user buying behavior or ratings in the past. Item-item-based filtering is based on common items (products) of similar users.

Association rule mining has been a popular technique for marketing professionals for quite some time now, often referred to as market basket analysis. Whenever the nature of data is categorical or non-numeric, association rule mining becomes crucial to underlying mining patterns in data transactions. This method identifies the frequent co-occurrence of data points in large databases using metrics such as support and confidence for the transaction. Support provides information about the frequency of the data co-occurrence, whereas confidence looks at all the transactions to identify the co-occurrence rate. If a relationship surpasses the threshold support and the threshold confidence, the relationship between the items is considered a valid rule that can be used to frame policies in the future (bundling of offers, store planogram, etc.).

Probabilistic Topic Models

Probabilistic topic models are a suite of ML algorithms that can decipher information from a large amount of textual data to capture the themes and the inter-relations among the topics (Blei, 2012). In simpler terms, topic models enable users to identify the topics in a large document using probabilistic word scores. Topic modeling is an unsupervised learning approach and hence does not require the pre-labeling of data. Probabilistic topic models have been extensively applied in text mining applications for marketing (Jacobs et al., 2016, 2021). The advantage of using the topic models is that it renders a large amount of data to be scalable without substantial loss of information. Scholars have used topic models for user personalization in digital content marketing (Gregoriades et al., 2021), ethnic marketing (Moro et al., 2019), Tourism (Christodoulou et al., 2020), and Airbnb reviews (Ding et al., 2020). Latent Dirichlet Allocation (LDA) is the simplest and most popular topic modeling technique. LDA assumes that each document comprises a pre-decided number of topics. The algorithm first assumes a distribution over the topics. Then, for each of the words, a topic is assigned. This gives rise to a certain number of words allocated to the pre-defined topics. Based on the words allocated to each of the topics, topic names are assigned. For instance, in a certain monetary policy document, words such as price, rise, inflationary, increase, and economy are gathered through LDA. The topic could be named inflation by the analyst. Hence, collectively the document could be made up of the defined topics. Topic models can be used to identify customer purchase behavior. Individual customer feedback can be considered a document distributed over various topics. The scores for each topic could then be used to predict buying propensity of new customers based on historical data. Hence, the individual buying behavior of customers could be recognized using the model. Topic modeling can also find applications in AI-based chatbots. The topics discussed could be analyzed to develop a customer engagement score to understand and predict the likelihood of a customer using the chat application in the future.

Customer Experience Management

Customer experience can be defined as a customer's response toward an organization while interacting with the organization (Kranzbühler et al., 2018). The response is multi-faceted, whether before, during, or after the interaction and over time. Although the arrival of big data and advanced ML techniques has resulted in greater insights from the data, the insights might not always result in a better customer experience. Organizations need specific strategies for data analysis to ensure better customer experiences. Data availability can improve customer experience. However, it's important to identify what kind of data can help improve it (Holmlund et al., 2020). For this, organizations must first define what customer experience entails for them and their customers. Then, identify the varied data sources directly impacting customer touchpoints. In the short term, the organization may focus on a couple of touch points. In the long term, organizations can strategize to identify the most crucial touchpoints across the customer's journey (Homburg et al., 2017). Firms should also consider the cost-benefit of implementing changes at the touch points by conducting a feasibility analysis for customer experience implementation.

Methodology Adopted for Review

We have used the integrative method to analyze the literature, which is considered an appropriate tool for the review of new and emerging topics (Snyder, 2019). We develop a conceptual framework to illustrate the role of data in AI for personalization. We explain the constructs we have used in this section as we culminate toward a holistic theoretical view of data and its significance for personalization.

Results

We summarize the theoretical framework presented in the methodology section. We present a multidimensional conceptualization of the role of data in AI-based personalization, and the framework can be divided into four distinct layers. The first layer conceptualizes the identification of data using customer activities on the internet. It is a part where we can arrive at the business problem and identify data and features which can be useful in exploring customer behavior. Data at this stage is raw and unstructured and not appropriate for analysis. The data must be stored in a big data environment within data lakes or in the organization's internal databases.

Data quantity and quality are investigated in layer 2 of the framework. Data quality is a multidimensional construct (Ballou et al., 1998; Ballou & Pazer, 1985; Pipino et al., 2002). Organizations must assess quality from individual subjective opinions on the data and objective valuation without biases at this stage. Although it might appear counterintuitive, subjective assessments are crucial to ensure that the stakeholders of the data share a similar understanding of the usefulness of the data for the identified task. The data custodians and the data users may have different opinions on data quality, often resulting in inefficient systems. The objective assessment of data can be of two types, task-independent, and task-dependent (Pipino et al., 2002). As the name suggests, task-independent assessments do not consider the context of the data collected. They are data agnostic and apply to datasets in general. On the other hand, the task-dependent assessment considers the business context, rules, database constraints, task feasibility, and other rules and regulations. Once the objective and subjective data assessment is complete, the data could be enhanced further by either discovery, augmentation, or generation. These three methods for data collection have been explained in the methodology section.

In its current raw form, data may contain multiple redundancies and inconsistencies to apply ML-based algorithms directly. In layer 3, data pre-processing ensures that the data is converted into a usable form for meaningful data analysis. Data pre-processing can be roughly divided into two major techniques: data preparation and data reduction (García et al., 2016). The different data preparation and reduction techniques are demonstrated in Fig. 7.2.

Data preparation entails modifying the data to make it suitable for the application of ML; it consists of changing unstructured data structured form, missing value treatment, combining data from different sources, and normalizing data to ensure easier interpretation of results. On the other hand, data reduction is concerned with reducing data complexity to ensure efficiency in algorithm run time and space requirements and parsimonious models. Feature selection removes any redundant data from the model, whereas instance selection techniques try to remove the data without compromising the dataset's quality. Both these methods are

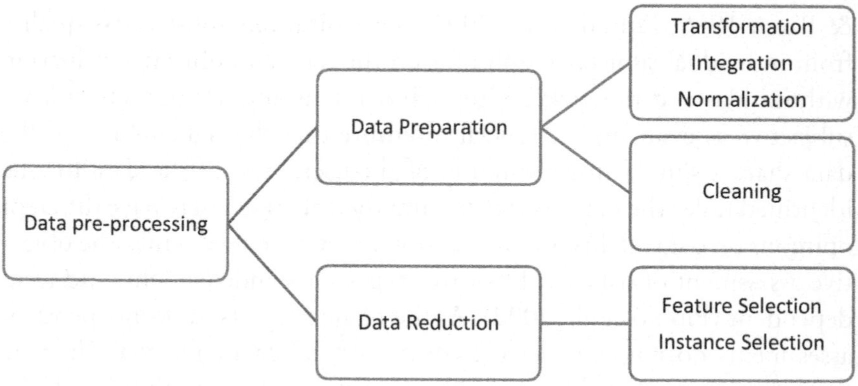

Fig. 7.2 Break down of data pre-processing (adapted from Garcia (2016))

extremely useful when dealing with large complex datasets. The dataset is now ready for analysis.

Finally, layer 4 of the data entails applying ML/DL techniques to the datasets, analyzing the results, and applying the results in real-life scenarios. It is important to emphasize that it takes multiple instances of trial and error to get to the threshold efficiency needed for the results to be applied in real-life scenarios. Therefore, multiple metrics, such as accuracy, precision, recall, F-score, and so on, could be used here for the minimum threshold of efficiency to be agreed upon by all the stakeholders. Once the model is live, it can be further enriched by adding customer data based on the prediction performance of the algorithm.

Implications

As human–computer interaction becomes more pervasive with every passing day, the requirement for personalization will become even more conspicuous shortly. Researchers and practitioners can explore various open areas in AI-based personalization. Profound among them is the identification of sarcasm in customer feedback. Personalization often requires scalability, which can be achieved through dimensionality reduction. While dealing with textual data in cases of emotion detection in chat or topic identification in probabilistic modeling methods, the danger of misrepresentation of customer sentiment exists. Hence, identifying user emotion is an open area for research with massive potential to interpret customer needs and appropriately respond to those correctly. There is also a need to understand and quantify customer satisfaction over a longer period. Once the algorithms have been applied to the data for recommendations, they might return decent recommendations in the short term. However, the recommendations might become ineffectual due to the implicit assumption that customer liking remains the same over time. Therefore, marketing experts must devise ways to capture a customer satisfaction score for recommendations over a longer period. A comparative study between personalized and generalized recommendation strategies could also help identify factors that lead to customer satisfaction across the strategies. This could help organizations develop concrete and actionable policies to ensure customer retention.

There has been a steady growth in the application of data analytics for social media data for marketing and personalization. However, only a few research studies focus on sustainability at a personalized level (Ilieva & McPhearson, 2018). Social media data could significantly improve social equity, public health, and environmental research. This is a research area with multiple open questions. Individual responses to natural disasters, urban settlements, depression, neighborhoods, and crimes are economic facets that researchers could look at using social media data. Environmental facets such as community perception of air quality, water quality, and access to parks and open access areas could also be studied. Furthermore, social equity issues such as population segregation and the evolution of protests can be researched in depth using data available on social media. Much potential is yet to be explored in sustainability using individual-level data. The area of sustainability naturally leads us to open responsible data science, which is again an open area for researchers to explore. Stakeholders today are increasingly suspicious of intrusive data usage by organizations. Automated decision-making could have inherent biases, and data confidentiality measures could be misused with a lack of transparency (van der Aalst et al., 2017). To address these problems, responsible data science practices need to be integrated into the design phase of data science solutions. Responsible data science revolves around four critical questions in literature (Van der Aalst, 2016): Fairness, accuracy, confidentiality, and transparency (FACT) (Responsible data science initiative, 2016). Fairness is to avoid prejudice or inherent bias in data science-based conclusions; accuracy entails avoiding any guesswork in data analysis to ensure a guaranteed level of performance; confidentiality ensures data is secure and not intrusive; and finally, transparency deals with indisputable methods and results clear all stakeholders. These responsible data science areas are at a nascent stage and have huge research potential for academicians.

Educating consumers about their right to privacy is also the responsibility of the organizations. Many organizations try to adopt conspicuous ways to track their customers through internet cookies and unwanted subscription emails. Several social media platforms have been accused of tracking customer activity on other websites without explicit permission from the user. These activities tend to erode customer trust and could be h harm the organization's reputation in the long run, leading to eventual loss of business and e. Therefore, it is incumbent upon organizations to maintain transparency in their policies to ensure customer engagement and participation. Thirdly, chatbot-based personalization is still in its nascent stages. Most of these chat applications are based on an API hitting the database and providing the answer to the customers' query without letting the customer experience true AI potential; this sometimes may lead to frustration and annoyance in customers. Organizations should try implementing ML and NLP-based solutions to their chat applications to give consumers true personalization. This would lead to a pleasant customer experience and cause better customer engagement, thus providing a competitive advantage in the long run.

Personalization using AI is a challenging task. It involves collecting a large amount of data and scaling it before applying ML-based learning for appropriate consumer recommendations. There is also a tacit assumption that user choices not changing with time once the algorithms have been deployed (Yoganarasimhan, 2017). The concern of loss of privacy of users is also a prominent issue. Users might not be comfortable with their data being used to recommend products (or services) and, as a result, might alter their browsing behavior rendering the personalization attempts futile. In addition to practical ML problems, personalization also brings ethical issues regarding users' right to privacy. In the future, researchers must attempt to understand the difference in user response between personalized and general or non-personalized recommendations through control group-based experiments.

A longitudinal study across multiple years could provide insights into how successful personalization strategies work at the ground level. This could also help us design responsible AI, which is non-intrusive and fair.

AI-based personalization is the next frontier for marketing experts. Although personalization has existed in various forms through the years, AI-based solutions profoundly impact its widespread adoption. Organizations have been able to reach and interact with a global audience at a level that would have been impossible even a decade back. This chapter focuses on data collection, processing, quality, and analysis methods that utilize state-of-the-art AI technologies. We created a conceptual framework for the role of data in AI-based personalization. We defined data types and collection techniques based on our conceptual framework. We also looked at how data analysis using advanced machine learning and deep learning techniques can aid in AI-based personalization. We discussed the profound role of recommender systems and probabilistic topic models in unearthing customer intentions and future buying behavior. Next, we touched upon the role of data science in the customer experience journey. We then discuss personalization's implications and way forward, focusing on ensuring customer transparency and information symmetry. Next, we conferred on the implications of personalization on academicians and practitioners. Finally, we discussed the need to identify the long-term implications of personalization through defined metrics.

The question of ethical AI is one of the most contentious issues when dealing with personalization. The rapid growth of AI has also led to the growth in the study of ethics in AI. However, regrettably, AI ethics are still nascent; thus, there are hardly any concrete ethical guidelines for firms to follow. The rise of ethical AI has, in a way, been triggered by the breach of user privacy rights. Academicians have a larger role to play in this context. The ideas present in behavioral ethics, social sciences, and philosophy need to be married to AI for new projects or research work. This could be a critical line of research in the future. Firms must ensure that technological achievements are not bereft of moral considerations.

Like other studies, this study, too, is not free from limitations; because this is a theoretical study, our reasoning and implications derive from past studies. Hence, we do not present any empirical evidence to test the framework. However, we hope that future researchers carry out survey-based empirical research on the framework to gather evidence and validate the study.

References

Abdulquadri, A., Kieu, T. A., & Nguyen, N. P. (2021). Digital transformation in financial services provision: Perspective to the adoption of chatbot. *Journal of Enterprising Communities: People and Places in the Global Economy, 15*(2), 258–281.

Ballou, D., Wang, R., Pazer, H., & Tayi, G. K. (1998). Modeling information manufacturing systems to determine information product quality. *Management Science, 44*(4), 462–484.

Ballou, D. P., & Pazer, H. L. (1985). Modeling data and process quality in multi-input, multi-output information systems. *Management Science, 31*(2), 150–162.

Bernardi, L., Mavridis, T., & Estevez, P. (2019, July). 150 successful machine learning models: 6 lessons learned at booking.com. In *Proceedings of the 25th ACM SIGKDD international conference on knowledge discovery & data mining* (pp. 1743–1751).

Bhardwaj, A., Karger, D., Subramanyam, H., Deshpande, A., Madden, S., Wu, E., … & Zhang, R. (2015, August). Collaborative data analytics with DataHub. In *Proceedings of the VLDB endowment international conference on very large data bases* (Vol. 8, No. 12, p. 1916). NIH Public Access.

Blei, D. M. (2012). Probabilistic topic models. *Communications of the ACM, 55*(4), 77–84.

Blei, D. M., Ng, A. Y., & Jordan, M. I. (2003). Latent Dirichlet allocation. *Journal of Machine Learning Research, 3*(Jan), 993–1022.

Caruana, R., & Niculescu-Mizil, A. (2006, June). An empirical comparison of supervised learning algorithms. In *Proceedings of the 23rd international conference on Machine learning* (pp. 161–168).

Christodoulou, E., Gregoriades, A., Pampaka, M., & Herodotou, H. (2020, June). Combination of topic modelling and decision tree classification for tourist destination marketing. In *International conference on advanced information systems engineering* (pp. 95–108). Cham: Springer.

Couper, M. P. (2017). New developments in survey data collection. *Annual Review of Sociology, 43*, 121–145.

Daniel, F., Kucherbaev, P., Cappiello, C., Benatallah, B., & Allahbakhsh, M. (2018). Quality control in crowdsourcing: A survey of quality attributes, assessment techniques, and assurance actions. *ACM Computing Surveys (CSUR), 51*(1), 1–40.

Davenport, T., Guha, A., Grewal, D., & Bressgott, T. (2020). How artificial intelligence will change the future of marketing. *Journal of the Academy of Marketing Science, 48*(1), 24–42.
Ding, K., Choo, W. C., Ng, K. Y., & Ng, S. I. (2020). Employing structural topic modelling to explore perceived service quality attributes in Airbnb accommodation. *International Journal of Hospitality Management, 91*, 102676.
Donmez, P., Lebanon, G., & Balasubramanian, K. (2010). Unsupervised supervised learning I: Estimating classification and regression errors without labels. *Journal of Machine Learning Research, 11*(4), 1323.
Ekstrand, M. D., Riedl, J. T., & Konstan, J. A. (2011). Collaborative filtering recommender systems. *Foundations and Trends® in Human-Computer Interaction, 4*(2), 81–173.
Farzindar, A., & Inkpen, D. (2015). Natural language processing for social media. *Synthesis Lectures on Human Language Technologies, 8*(2), 1–166.
García, S., Ramírez-Gallego, S., Luengo, J., Benítez, J. M., & Herrera, F. (2016). Big data pre-processing: Methods and prospects. *Big Data Analytics, 1*(1), 1–22.
Goldberg, Y., & Levy, O. (2014). word2vec explained: Deriving Mikolov et al.'s negative-sampling word-embedding method. *arXiv preprint arXiv: 1402.3722.*
Goldenberg, D., Kofman, K., Albert, J., Mizrachi, S., Horowitz, A., & Teinemaa, I. (2021, March). Personalisation in practice: Methods and applications. In *Proceedings of the 14th ACM international conference on web search and data mining* (pp. 1123–1126).
Gregoriades, A., Pampaka, M., Herodotou, H., & Christodoulou, E. (2021). Supporting digital content marketing and messaging through topic modelling and decision trees. *Expert Systems with Applications, 184*, 115546.
Haenlein, M., & Kaplan, A. (2019). A brief history of artificial intelligence: On the past, present, and future of artificial intelligence. *California Management Review, 61*(4), 5–14.
Halevy, A., Korn, F., Noy, N. F., Olston, C., Polyzotis, N., Roy, S., & Whang, S. E. (2016, June). Goods: Organizing Google's datasets. In *Proceedings of the 2016 international conference on management of data* (pp. 795–806).
Holmlund, M., Van Vaerenbergh, Y., Ciuchita, R., Ravald, A., Sarantopoulos, P., Ordenes, F. V., & Zaki, M. (2020). Customer experience management in the age of big data analytics: A strategic framework. *Journal of Business Research, 116*, 356–365.

Homburg, C., Jozić, D., & Kuehnl, C. (2017). Customer experience management: Toward implementing an evolving marketing concept. *Journal of the Academy of Marketing Science, 45*(3), 377–401.

Ilieva, R. T., & McPhearson, T. (2018). Social-media data for urban sustainability. *Nature Sustainability, 1*(10), 553–565.

Jacobs, B., Fok, D., & Donkers, B. (2021). Understanding large-scale dynamic purchase behavior. *Marketing Science, 40*(5), 844–870.

Jacobs, B. J., Donkers, B., & Fok, D. (2016). Model-based purchase predictions for large assortments. *Marketing Science, 35*(3), 389–404.

Khrais, L. T. (2020). Role of artificial intelligence in shaping consumer demand in E-commerce. *Future Internet, 12*(12), 226.

Kim, Y. S., & Yum, B. J. (2011). Recommender system based on click stream data using association rule mining. *Expert Systems with Applications, 38*(10), 13320–13327.

Kranzbühler, A. M., Kleijnen, M. H., Morgan, R. E., & Teerling, M. (2018). The multilevel nature of customer experience research: An integrative review and research agenda. *International Journal of Management Reviews, 20*(2), 433–456.

Kumar, R., Mukherjee, S., Choi, T. M., & Dhamotharan, L. (2022). Mining voices from self-expressed messages on social-media: Diagnostics of mental distress during COVID-19. *Decision Support Systems, 113792*, 113792.

LeCun, Y., Bengio, Y., & Hinton, G. (2015). Deep learning. *Nature, 521*(7553), 436–444.

Lops, P., Gemmis, M. D., & Semeraro, G. (2011). Content-based recommender systems: State of the art and trends. In F. Ricci, L. Rokach, & B. Shapira (Eds.), *Recommender systems handbook* (pp. 73–105). Springer-Verlag.

Mikolov, T., Sutskever, I., Chen, K., Corrado, G. S., & Dean, J. (2013). Distributed representations of words and phrases and their compositionality. *Advances in Neural Information Processing Systems, 2*, 26.

Mogaji, E., & Nguyen, P. N. (2022). Managers' understanding of artificial intelligence in relation to marketing financial services: Insights from a cross-country study. *International Journal of Bank Marketing, 40*(6), 1272–1298.

Mogaji, E., Soetan, T., & Kieu, T. (2020). The implications of artificial intelligence on the digital marketing of financial services to vulnerable customers. *Australasian Marketing Journal, 29*(3), 235–242.

Moro, S., Pires, G., Rita, P., & Cortez, P. (2019). A text mining and topic modelling perspective of ethnic marketing research. *Journal of Business Research, 103*, 275–285.

Mukherjee, S., & Bala, P. K. (2016). Gender classification of microblog text based on authorial style. *Information Systems and e-Business Management, 15*(1), 117–138.

Mukherjee, S., & Bala, P. K. (2017). Sarcasm detection in microblogs using Naïve Bayes and fuzzy clustering. *Technology in Society, 48*, 19–27.

Nguyen, T. H., & Grishman, R. (2013, November). Modeling skip-grams for event detection with convolutional neural networks. In *Proceedings of the 2016 conference on empirical methods in natural language processing* (pp. 886–891).

Pipino, L. L., Lee, Y. W., & Wang, R. Y. (2002). Data quality assessment. *Communications of the ACM, 45*(4), 211–218.

Rafieian, O., & Yoganarasimhan, H. (2022). AI and personalisation. Available at SSRN 4123356.

Ramaswamy, S., & DeClerck, N. (2018). Customer perception analysis using deep learning and NLP. *Procedia Computer Science, 140*, 170–178.

Responsible Data Science Initiative. (2016). *Responsible data science*. http://www.responsibledatascience.org/.

Ricci, F., Rokach, L., & Shapira, B. (2011). Introduction to recommender systems handbook. In F. Ricci, L. Rokach, & B. Shapira (Eds.), *Recommender systems handbook* (pp. 1–35). Springer.

Roh, Y., Heo, G., & Whang, S. E. (2019). A survey on data collection for machine learning: A big data-ai integration perspective. *IEEE Transactions on Knowledge and Data Engineering, 33*(4), 1328–1347.

Sagiroglu, S., & Sinanc, D. (2013, May). Big data: A review. In *2013 international conference on collaboration technologies and systems (CTS)* (pp. 42–47). IEEE.

Schafer, J. B., Frankowski, D., Herlocker, J., & Sen, S. (2007). Collaborative filtering recommender systems. In *The adaptive web* (pp. 291–324). Springer.

Seljan, S., & Dunđer, I. (2014). Combined automatic speech recognition and machine translation in business correspondence domain for English-Croatian. *International Journal of Industrial and Systems Engineering, 8*(11), 1980–1986.

Singhal, A., Sinha, P., & Pant, R. (2017). Use of deep learning in modern recommendation system: A summary of recent works. *arXiv preprint arXiv*: 1712.07525.

Snyder, H. (2019). Literature review as a research methodology: An overview and guidelines. *Journal of Business Research, 104*, 333–339.

Terrizzano, I. G., Schwarz, P. M., Roth, M., & Colino, J. E. (2015, January). Data wrangling: The challenging journey from the wild to the lake. In *CIDR*.

Thorat, P. B., Goudar, R. M., & Barve, S. (2015). Survey on collaborative filtering, content-based filtering, and hybrid recommendation system. *International Journal of Computer Applications, 110*(4), 31–36.

Van den Oord, A., Dieleman, S., & Schrauwen, B. (2013). Deep content-based music recommendation. *Advances in neural information processing systems*, 26.

Van der Aalst, W. M. (2016). Green Data Science. *ICEIS, 2016*, 9.

van der Aalst, W. M., Bichler, M., & Heinzl, A. (2017). Responsible data science. *Business & Information Systems Engineering, 59*(5), 311–313.

Van Meteren, R., & Van Someren, M. (2000, May). Using content-based filtering for recommendation. In *Proceedings of the machine learning in the new information age: MLnet/ECML2000 workshop* (Vol. 30, pp. 47–56).

Verhagen, T., Van Nes, J., Feldberg, F., & Van Dolen, W. (2014). Virtual customer service agents: Using social presence and personalisation to shape online service encounters. *Journal of Computer-Mediated Communication, 19*(3), 529–545.

Verma, S., Sharma, R., Deb, S., & Maitra, D. (2021). Artificial intelligence in marketing: Systematic review and future research direction. *International Journal of Information Management Data Insights, 1*(1), 100002.

Wang, X., & Wang, Y. (2014, November). Improving content-based and hybrid music recommendation using deep learning. In *Proceedings of the 22nd ACM international conference on Multimedia* (pp. 627–636).

Wang, Y., Wu, X., & Hu, D. (2016, March). Using Randomized Response for Differential Privacy Preserving Data Collection. In *EDBT/ICDT Workshops* (Vol. 1558, pp. 0090–6778).

Wirth, R., & Hipp, J. (2000, April). CRISP-DM: Towards a standard process model for data mining. In *Proceedings of the 4th international conference on the practical applications of knowledge discovery and data mining* (Vol. 1, pp. 29–39).

Yoganarasimhan, H. (2017). Identifying the presence and cause of fashion cycles in data. *Journal of Marketing Research, 54*(1), 5–26.

Zhu, X. J. (2005). Semi-supervised learning literature survey.

Theme III

Customer Experience

8

Artificial Intelligence for Seamless Experience Across Channels

Nguyen Phong Nguyen and Emmanuel Mogaji

Introduction

The increased adoption of multi-channel retailing has emphasized the need for seamless experiences across different channels (Batra, 2019). In this digital era, many business entities are utilizing different channels to sell their merchandise, including online stores, offline stores, mobile apps, and social media (Voorveld et al., 2016). The effective operation of these multiple retailing channels is anchored on delivering better customer experiences on all available platforms (Batra, 2017). Seamless

N. P. Nguyen
University of Economics Ho Chi Minh City, Ho Chi Minh City, Vietnam
e-mail: nguyenphongnguyen@ueh.edu.vn

E. Mogaji (✉)
University of Greenwich, London, UK

Department of Research Administration and International Relations,
University of Economics Ho Chi Minh City, Ho Chi Minh City, Vietnam
e-mail: E.O.Mogaji@greenwich.ac.uk

© The Author(s), under exclusive license to Springer Nature Switzerland AG 2023

J. N. Sheth et al. (eds.), *Artificial Intelligence in Customer Service*,
https://doi.org/10.1007/978-3-031-33898-4_8

customer experience is a significant consideration in the current technological era. It is anchored on the increasing proliferation of on-demand services and same-day deliveries (Steinhoff et al., 2018).

Artificial intelligence (AI) enables seamless customer experiences in a multi-channel retailing environment (Wilson & Daugherty, 2018). AI is a non-human intelligence encompassing a set of algorithms, machines, and systems that demonstrate intelligence (Shankar, 2018). It enables machines to constantly learn from the environment to obtain knowledge and develop human-like cognitive skills (Nair, 2021). AI is essential in delivering a seamless experience in multi-channel retailing through its capabilities to analyze and interpret large datasets and utilize insights to predict prospects (Shankar, 2018). In addition, it helps ease human effort and accommodates the inability to perform certain complex tasks to enable automation across different business areas (Jayakrishnan, 2022). Therefore, AI has been the most suitable technological integration for businesses that want to deliver the best customer experience across different channels.

With the growing amount of data and computer processing capabilities (Dwivedi et al., 2021), it is essential to explore how artificial intelligence (AI) and machine learning (ML) can be integrated with the data to enhance business operations. This information about adopting AI for a seamless experience across the various business channels would be useful for key stakeholders. First, business owners are contemplating their digital transformation strategies. These business owners and brands could be financial service providers, media and entertainment companies, and higher education institutions exploring how to gain insights from their data and what they can do with those insights (Dwivedi et al., 2021; Mogaji & Nguyen, 2022a). Second, this insight would be relevant for AI developers working on a different algorithm for business and developing their platform for individuals and businesses (Mogaji et al., 2023). This could be a tech company working with a university to develop learning analytics or developing its platform to curate student learning materials (Sharma et al., 2022). Third, policymakers need to be aware of the needs of businesses and customers, ensuring an enabling environment to support digital transformation (Dwivedi et al., 2021).

This chapter considers how AI can be harnessed to deliver seamless customer experiences through its technological integrations. To achieve this aim, the study adopts an integrative literature review approach to critically review and synthesize the literature on seamless customer experiences (CX), AI for omni-channel retail experience, AI for personalized engagement marketing (PEM), and machine learning (ML) for multi-channel messaging applications, which enables us to develop a new theoretical framework and perspective (Snyder, 2019) to enhance the better understanding of how to use artificial intelligence to enhance the customer experience across the channels. In doing this, the study contributes to the existing body of work on data-driven decision-making, business analytics, and data-driven customer service (Ahn et al., 2022; Dwivedi et al., 2021; Kim et al., 2023; Khalil & Abdelli, 2022), with theoretical implications for academic researchers, managers, and policymakers exploring marketing prospects, digital transformation, and service quality in a data-driven business environment.

This section summarizes the literature on seamless customer experiences, AI for omnichannel retail experience, AI for personalized engagement marketing , and machine learning for multi-channel messaging applications. Recommendations and practical implications for the stakeholders subsequently follow this section. The chapter concludes with a concluding remark highlighting the study's limitations and the agenda for future research.

Literature Review

Seamless Customer Experiences

Seamless customer experience is derived from the general concept of customer experience, which denotes the customer's experience with the products along their purchase journey (Nair, 2021). AI has been a major digital transformation for implementing seamless customer experiences by enhancing existing applications, uncovering hidden insights, establishing trends, and scaling the business' expertise (Sattarapu et al., 2023).

Many enterprises are adopting AI by deploying machine learning (ML), natural language processing (NLP), virtual agents, and predictive analytics as suitable ways to optimize customer outcomes and enhance customer interactions (Kohler et al., 2023). The use of AI in enhancing customer experiences is meant to increase the effectiveness and efficiency of customer experiences by enabling personal experiences that could not be easily implementable without this technology (Wynne & Lyons, 2018). It makes it easier to personalize and address individual and personal concerns faster and more efficiently than humans can. A seamless customer experience can be implemented using AI through target messaging, online support services, targeted advertising, personalized content creation, intelligent customer insights, and churn reduction (Kaliraj & Devi, 2021). In the era of AI, a seamless customer experience is a major competitive advantage for any business. AI incorporates intelligent experience engines utilizing customer data to deliver high-quality end-to-end customer experiences. The necessity of AI in enhancing customer experiences is also anchored on the need to scale up the personalization of customer interactions, which humans cannot achieve (Russell & Norvig, 2016). It requires omni-channel integrations to merge the flow of the customer's digital and personal experiences (Von Briel, 2018). AI can be deployed to build intelligent experience engines that will focus on combining the micro goals of the customer journey for enhanced interactions (Rustholllkarhu et al., 2022). It also makes it possible to connect user data with relevant insights to enable various marketing initiatives, such as geographical and gender-based targeting. Personalization across numerous channels helps deliver offers based on the individuals' behaviors and preferences (Andersson et al., 2017). Investments in AI-powered data analytics are also vital in personalizing customer experiences to their context to enhance the appeal. It is a great intervention in the current era, where customers value quick responses and the ability to communicate anytime.

A seamless customer experience is a central necessity in the customer purchase journey, and it integrates with the increasing significance of sensory marketing (Lemon & Verhoef, 2016). Sensory marketing uses the consumers' senses to impact their judgment, perception, and behavior (Krishna, 2013). Therefore, technology-based service delivery systems have become a key part of commerce and significantly impact the overall customer experience (Pelet et al., 2019). AI-based resource management and topology will help realize seamless customer experiences by efficiently adjusting resource utilization depending on the changes in the environments and dynamics in customer expectations (Zhang & Zhu, 2020). This functionality is attained through harnessing dispersed computing resources and devices to circumvent individual devices' storage and power limitations. In addition, AI technologies can be harnessed to enhance the customer experience through ML and NLP in chatbots, enhancing seamless integration across devices and providing personalized offers using analytics (Biswas et al., 2020).

Bilgeri (2020) established that AI could be harnessed for customer relationship management (CRM) and CX management. It can be deployed through various integrations such as customer service, process automation, lead prediction and profiling, pricing, content creation, and conversational commerce. AI helps in enhancing efficiency and productivity but can also be utilized to provide better services to customers, which creates added value (Kunz et al., 2019). For example, automated customer service involves using AI and its associated technologies to power virtual assistants that carry out tasks such as handling calls and other customer queries. In content creation, digital data is utilized for automated content generation and customizing content using Natural Language Generation (NLG) (Paluch & Wirtz, 2020).

AI for Omni-Channel Retail Experience

The omni-channel retail experience encompasses creating a streamlined and connected customer experience across offline and online platforms (Narayanan & Nandagopal, 2016). This approach enables businesses to enhance their sales practices through flexible customer engagements depending on their preferences. An omni-channel retail experience also aids in providing a seamless experience across multiple channels to save time on routine tasks (Li et al., 2019). The necessity for an omni-channel retail experience is anchored on the increasing perception that customers want a flawless experience across different channels. It provides a competitive advantage, enhances customer satisfaction, and promotes better inventory planning. According to Cai and Lo (2020), Omni-channel retailing has become a common strategy in the new era characterized by digitization, big data, social media, and other emerging technologies. Verhoef (2021) regard omni-channel management as the synergetic management of the available channels and customer touchpoints, optimizing the performance and customer experience across channels. Omni-channel retailing management is anchored on the significance of cross-channel operations such as cross-channel pricing, product promotions, marketing research, and customer communication (Gabisch & Gwebu, 2011). According to Cao and Li (2015, cross-channel integration significantly impacts sales growth among retailers. Shen et al. (2018) further indicated that the quality of cross-channel integration greatly impacts the overall user experience. The major considerations in cross-channel integration include the variety of channel choices and content, transparency of the service, and consistency in the processes. Omni-channel retailing mainly targets integrating all channels and optimizes overall sales and technology (Bell et al., 2014; Brynjolfsson et al., 2013). AI plays a central role in omni-channel retailing management since identity attractiveness, uncertainty, and switching costs determine the customer experience in this retailing environment (Li et al., 2019). It helps implement digital strategies such as gathering digital data, creating a digital brand and digitizing the systems for effective cross-channel integration (Frishammar et al., 2018).

According to Grewal et al. (2018), the retail sector is evolving thanks to the increased proliferation of mobile phones and smart devices. Bădică and Mitucă (2021) support the retailing evolution due to the development of digital technologies and computer interconnection via the internet, which have paved the way for the evolution from traditional marketing to digital marketing. AI is a significant contributor to the retailing evolution in combination with other current technologies, including ML, robotics, the internet of things (IoT), social media, and blockchain techno, enabling various retailing aspects such as retailing analytics (Bradlow et al., 2017). Retailing analytics entails one of the most significant functions of AI since it enables retailers to gain better insights into their customers. These insights are derived from various datasets, including location, purchase, and social media data, which can be harnessed for value creation (McAfee et al., 2012). Many retailers are investing in retail analytics in place of traditional loyalty programs to enhance system integration and make effective use of available datasets to enhance consumer engagement (Grewal et al., 2018). NLP has also been harnessed as an essential component of retailing analytics, whose results are used to facilitate managerial decision-making (Kumar & Saha, 2017). Other emerging technologies, such as big data, enable AI integrations in retail. It enables retailers to collect essential data from enterprise systems, demographic information, and social media usage statistics. Retailers can uncover unique insights using this data alongside customer loyalty data and demographic and geographic information to enhance the customer experience (Hallikainen et al., 2020). Another important AI integration in the retailing landscape involves using frontline service robots (FSR), which will be deployed in fulfillment centers (Wirtz et al., 2018). In addition, drones and driverless vehicles have been suggested as potential replacements for human labor, enhancing the overall customer experience (Kim et al., 2023).

Modern AI applications are linked to various advanced data processing technologies to perform functions such as solving problems, learning, reasoning, planning, and communicating (Paschen et al., 2020). ML refers to such capabilities whose methods include decision trees, artificial neural networks, random forests, and regression (Asare-Frempong & Jayabalan, 2017). ML obtains patterns from data using different ML methods, including NLP and image recognition (Mogaji et al., 2020). AI is the future of e-commerce, where it will empower various technology-driven intelligent processes to enhance their smartness, speed, and levels of personalization (Raisch & Krakowski, 2020). The increasing utilization of digital tools for customer interactions implies that AI offers businesses new opportunities to manage these interactions within digital spaces. AI technologies leverage the data-generative nature of e-commerce environments to transform and enhance sales and marketing processes (Wirtz et al., 2018). The major roles of AI revolve around its management functionalities, including the transformation of task management, automated accounting management, and robotization of customer service (Korhonen et al., 2021). AI also plays a crucial role in customer journeys to enable effective experience management (Koohang et al., 2023).

According to Chi et al. (2019), a combination of AI and IoT can be deployed to enhance the Omni-channel retail experience through the scope of the drama theory. The main focus is creating a single business platform with unified back-end and front-end systems for a better customer experience (Chen & Chen, 2010). The drama theory is deployed in the retailing service industry to create a service interaction process that resembles a theatre performance. Under this theory, the main assumption is that the four major elements of theatre performance, including actors, setting, performance, and audience, significantly impact customer satisfaction (Kossakowski et al., 2018). AI and IoT enable the development of such a scene by exploring the core needs and behaviors of the consumers and integrating this data into the service to better their experience. Furthermore, the same information can also be utilized to develop an immersive experience across channels by targeting the specific advantages of each channel. Bae and Wooldridge (2020) further established that AI is an essential component of omni-channel retailing as it helps in collecting consumer data, making product recommendations, enhancing visual merchandising, and providing consumer insights.

AI for Personalized Engagement Marketing

AI is a significant innovation that will help deliver value to users, especially through PEM among retailers (Kumar et al., 2019). AI is actively shaping many aspects of human lifestyle through personalization. Personalization and AI converge in the marketing sphere as it helps collect customer data and utilize it to determine the suitable marketing mix for a given individual (Jordan & Mitchell, 2015). The personalization process is relevant in engagement marketing because customer relationships are characterized by emotional bonding progress toward a state of engagement. Therefore, personalization initiates positive relationships, significantly affecting customer engagement (CE) behaviors (Mogaji et al., 2020). Since engagement encompasses the customers' behaviors, attitudes, and level of connectedness, various AI-related integrations can be deployed to enhance this outcome. AI enables exceptionally high levels of personalization through its deep learning (DL) and data-driven integrations (Chiu & Lin, 2018). Its unobtrusive nature helps create an endearing bond with the customers by working on a personal level. This functionality can be harnessed for customer value creation. AI-powered tools are also deployed in digital curation for PEM, where the main focus is on the automatic selection of products, advertising messages, web content, and dynamic pricing (Karp, 2016). It deploys customer transactions, communication, and consumption patterns data to predict purchase preferences. ML algorithms such as DL, collaborative filtering, k-nearest neighbors, and unsupervised clustering have been harnessed for such functionalities (Mullainathan & Spiess, 2017). Self-learning neural networks have also become a major revolution in PEM. This development combines NLP technologies to generate suitable customer buying options and help in buyer decision-making and implementation. AI enables PEM by suggesting a wide range of recommended products with different configurations for easier decision-making (Pansari & Kumar, 2017). It also makes comparing similar products and prices easier through continuous knowledge organization, management, and sophisticated customer information processing mechanisms. AI is increasingly used in marketing, whereby the tools and devices deployed extensively rely on consumer behavior. It is deployed in marketing decision-making to solve many traditional and digital problems.

Machine Learning for Multi-channel Messaging Applications

Multi-channel messaging frameworks utilizing ML provide intelligent customer communication pathways (Salami & Mnkandla, 2021). These communication pathways enhance seamless multi-channel integration on a single retailing platform. The ability to relay the appropriate information at the right time and use the most suitable channel has the impetus for the success of any retailing venture. These messaging models can be built on existing communication frameworks, including email, SMS, and instant messaging services (Ma et al., 2019). The automation capabilities of AI/ML systems make it possible to provide virtual assistants for precise translation and interpretation of the human language (Bisoi et al., 2020). Smart AI chatbots constitute a similar integration whose major functions involve solving business tasks and making decisions on various business and personal activities alongside automated financial service applications (Shabbir & Anwer, 2018). Multi-channel messaging capabilities enable the business to achieve sophisticated customer segmentation and enhance the overall customer experience (Jiang et al., 2020). They also promote sales across different channels through their advertising and commerce support via the available integrated channels. Providing different channels to communicate with the customers enhances the customer's perception of the service offering. Providing communication channels that match the customer's preferences is an effective integration attained through ML analytics (Salami & Mtsweni, 2016). AI/ML applications have continuously improved in many facets and currently apply in many domains, including facial recognition, pattern recognition, speech recognition, and dynamic machine learning (Nasteski, 2017). The AI and dynamic machine learning capabilities are important in ensuring the efficient and effective selection of messaging channels based on the available resources and channel integration modules.

According to Panda and Chakravarty (2022), AI and its associated technologies, including ML, NLP, and DL, are fuelling the evolution of Intelligent Information Services (IIS). The IIS systems enable the creation, collection, and accumulation of data through AI-powered tools and devices to enable higher-level automation that will replace human cognitive capabilities in completing various tasks (Rosário, 2021). It has enabled technological tools that connect humans and machines through intelligence. This connection harnesses other integrations, such as IoT systems, to enable the development of AI chatbots (Pal et al., 2022). The combination of AI and NLP is utilized to develop intelligent virtual assistants (IVAs) implemented through smart AI chatbots (Balakrishnan et al., 2021). These chatbots simulate human conversations through audio or text and are integrated into various messaging applications, mobile apps, and websites (McNeal & Newyear, 2013). AI enables the operation of chatbots by discovering patterns in the available data to develop the intelligence required to perform tasks, manage information, and solve problems (Nuruzzaman & Hussain, 2018). IVA is essential to enhance customer experience since it facilitates an everyday user experience (UX) via voice or text-based channels.

Discussion

AI is significant in implementing seamless customer experiences because businesses can deploy it to enhance existing applications, establish trends, and uncover hidden insights from the available consumer data sets (Liu-Thompkins et al., 2022;Moore et al., 2022 ; Nguyen et al., 2022). Its associated technologies include ML and NLP alongside their associated integrations, such as predictive analytics and virtual agents. AI helps enhance customer experiences by optimizing their outcomes and enabling personalized experiences. Target messaging, target advertising, online support services, and personalized content are some of the AI-related functionalities that enable the implementation of seamless customer experiences.

This application could be in the banks within the financial sector that can analyze consumer data to provide personalized credit offers to customers, recognize fraud, and provide credit ratings and scores (Mogaji & Nguyen, 2022b; Mogaji et al., 2022; Nguyen et al., 2022). Adopting AI in retail services to enhance customer experience and shopping practices (Moore et al., 2022; Nguyen et al., 2022). Media and entertainment sectors like Netflix and Paramount can use analytics to know what type of movies people are watching, what time they are watching, and what interests them; this insight can be used in commissioning newer projects, offering recommendations, and also for advertising purposes (Fagerjord & Kueng, 2019; Hadida et al., 2021). Likewise, the universities can benefit from learning analytics to evaluate how well their students are engaging with the virtual learning environments (VLEs), their attendance, and visits to the library, which can be used in determining students who need assistance and even predict future grades (Jain et al., 2022; Sharma et al., 2022).

The firms can also connect user data and relevant insights for marketing activities. AI systems also enable adjustments in resource utilization based on environmental changes and dynamics in consumer preferences. Entities can also harness AI in CRM and CX through lead prediction, automated customer service, profiling, and dynamic pricing. An omni-channel retail experience ensures a seamless customer experience across multiple channels through the synergetic management of customer touchpoints and channels. The operations related to omni-channel retailing include cross-channel pricing, market research, product promotions, and customer communication. Cross-channel integration impairs customer experience through various content and channel choices, service transparency, and process consistency. AI plays a crucial role in omni-channel retailing by enabling digital data collection, system digitization, and the creation of digital brands.

AI and its associated technologies, including ML, IoT, and robotics, enable various aspects of digital retailing, including retailing analytics (Abdulquadri et al., 2021; Chemma et al., 2022). Retailing analytics is vital in delivering seamless customer experiences by providing insights into the customers that can be harnessed for value creation (Brooksworth et al., 2023). NLP is crucial in retailing analytics alongside Big Data and Robotics, deployed in data collection, uncovering insights from consumer data, and deployment in fulfillment centers such as FSR. ML deploys NLP and image recognition to discover patterns from consumer data which constitute the retailing analytics. These advanced functionalities make AI the future of e-commerce by empowering technology-driven intelligent processes. A combination of AI and IoT is useful in enhancing the Omni-channel retail experience by exploring consumer needs and behaviors and integrating the data to deliver better experiences. AI also helps in making relevant product recommendations and enhancing visual merchandising. It is extensively deployed in PEM through its DL and data-driven integrations, making it possible to personalize customer relationships. PEM is implemented through the automatic selection of products, web content, advertising messages, and dynamic pricing. It is made possible through ML algorithms such as DL, k-nearest neighbors, collaborative filtering, and unsupervised clustering. Self-learning neural networks combine with NLP to recommend products and aid buyer decision-making. ML is deployed in multi-channel messaging to provide intelligent communication pathways that enable a seamless experience across different channels. It makes it possible to relay the right information to the right customer at the right time and use the most suitable channel.

Automation capabilities of AI/ML systems enable the development of virtual assistants in the form of smart AI chatbots. Smart AI chatbots can solve business tasks, make decisions and provide automated financial services. Multi-channel messaging enhances the customer experience through effective segmentation and promoting sales across different channels. ML analytics make it possible to provide communication channels that match the customers' preferences and consider other relevant aspects such as channel integration modules and availability of resources. Such capabilities enhance the customers' overall perception of the service offering.

Managerial Implications

Business leaders should consider investing in AI-enabled value chains for customer engagement, knowledge and insight management, operations management, and inventory management. They should also consider deploying AI in offline and online stores to enable point-of-sale (POS) digitization and automated systems for self-checkout. Retailers need to fast-track the adoption of robotics technology in-store management to provide intelligent guidance to the customer, facilitate the ordering process and address customer issues. The viability of acquiring AI-powered humanoid robots should also be assessed for visual merchandising and category management. Business managers should also consider how they will use insights from AI from tracing the customer purchase journey to address challenges in this process.

Business leaders specializing in the business-to-business (B2B) sector should consider developing competent managerial skills and digital capabilities for managing complex buying and selling processes in this sector, which have gradually been digitized (Dwivedi et al., 2021; Mogaji et al., 2023). Furthermore, tracking customer interactions across different channels and touchpoints will require significant investments in marketing and sales operations fusion through available AI tools (Nguyen et al.,

2022). Therefore, considerate budgetary allocations for implementing AI and its associated technologies are vital because they will help unlock new possibilities for managing customer interactions in digital retail environments.

The findings on the significance of AI in delivering seamless customer experiences should be utilized to develop a suitable framework for implementing AI-empowered tools in customer journey management. Data on AI-related technologies, their usage, and their capabilities should enable business leaders to evaluate the level of AI deployment in their entities' sales and marketing activities. It should also help them propose suitable areas for development and procedures to be followed in the implementation to harness the possibilities availed by AI. Business leaders should also devise digital retailing strategies coherent with the requirements in multichannel retailing and grow the use of AI to deliver better customer experiences.

Other considerations include evaluating the link between different retailing channels and how AI can be deployed to make the experiences similar in all scenarios. It will involve harnessing the different capabilities of AI as they apply to their business model and customizing to the buyers' needs on each channel. Lastly, marketers should collaborate with AI experts to re-imagine the customer service process using AI and work toward enhancing operational flexibility, level of productivity, and personalization of the commodities.

Business leaders should devise policies that support consumer data protection since AI systems collect large amounts of data about the consumer journey. The main measures under these policies should revolve around fair business practices and information disclosures. Other areas to be considered revolve around fair use of consumer data drawn from retailing analytics to prevent exploitation. The management should also develop policies that support skills development for their AI teams and measures to curb AI's negative impacts, such as algorithmic bias and data exploitation. Additionally, there should be significant investments in AI-related research and development to enable business entities to stay ahead of the curve in terms of its use and implementation. Lastly, business managers should devise policies enabling entities to attract, develop and retain talent in AI and acquire suitable data and digital infrastructures that support its integration.

Conclusion

Multi-channel retailing has become a common trend, with many businesses opting for different channels to sell their merchandise. Its advent has been met with the increasing need to deliver seamless experiences across different channels. AI is fronted as one of the suitable solutions for delivering seamless experiences across multiple channels based on its numerous technological capabilities. Its technological capabilities are associated with various technologies, including ML, NLP, DL, NLG, and robotics. AI is useful in delivering seamless customer experiences through its applications in PEM, enabling omni-channel retailing, sensory marketing, and multi-channel messaging capabilities.

Predictive analytics and virtual agents is a major AI-powered solution that helps optimize customer outcomes and enhance interactions. Personalization is crucial in delivering seamless customer experiences because the services and product offerings are customized to the customer's behaviors and preferences, making them more appealing. AI is also vital in CRM and CX through automation, conversational commerce support, and creating added value. Its application in omni-channel retailing has been associated with its digital data gathering and branding capabilities that enable cross-channel integration. In PEM, AI powers much functionality through self-learning neural networks, NLP, ML, and DL to help consumer decision-making and suggest relevant products to customers. The increased significance of AI in delivering seamless customer experiences implies that business managers should consider investing in AI-enabled value chains and fast track the adoption of robotics in the retail sector. They should also evaluate the level of AI utilization in their firms and make suitable recommendations for enhancing its application. Policy concerns related to data protection, talent acquisition, research and development, and addressing the negative impacts of AI.

While effort has been made to synthesize existing literature on this matter, there are some limitations to this study, and readers should interpret the provided insights in the context of these limitations. First, the study was highly conceptual, thus opening an opportunity for future research to explore and empirically validate the insights provided in this chapter. Second, like any other conceptual paper (e.g. Ahn et al., 2022; Kim et al., 2023; Khalil & Abdelli, 2022), the study gave a holistic view of seamless customer experiences, AI for omni-channel retail experience, AI for personalised engagement marketing and machine learning for multi-channel messaging applications, and no data has been collected to validate the propositions empirically. Future studies can contextualize and validate the conceptual framework in business contexts. Notwithstanding, the study offers future research opportunities to understand better how to use artificial intelligence for a seamless experience across the channels. Like others that have provided these theoretical insights, this study will likely spur future research and expand our knowledge about the subject matter. Future studies must collect qualitative and quantitative data from the key stakeholders to understand their challenges and opportunities for using artificial intelligence to enhance the customer experience across the channels.

References

Abdulquadri, A., Mogaji, E., Kieu, T. A., & Nguyen, N. P. (2021). Digital transformation in financial services provision: A Nigerian perspective to the adoption of chatbot. *Journal of Enterprising Communities: People and Places in the Global Economy, 15*(2), 258–281.

Ahn, M., Brohan, A., Brown, N., Chebotar, Y., Cortes, O., David, B., … & Zeng, A. (2022). *Do as I can, not as I say: Grounding language in robotic affordances.* arXiv preprint arXiv:2204.01691.

Andersson, T., Boedeker, M., & Vuori, V. (2017). Emotion-gauge: Analysing affective experiences in B2B customer journeys. In A. Kavoura, D. P. Sakas, & P. Tomaras (Eds.), *Strategic innovative marketing* (pp. 31–36). Springer. https://doi.org/10.1007/978-3-319-56288-9_5

Asare-Frempong, J., & Jayabalan, M. (2017). Exploring the impact of big data in healthcare and techniques in preserving patients' privacy. *IJCSNS, 17*(8), 143.

Bădică, A., & Mitucă, M. (2021). IOT-enhanced digital marketing conceptual framework. *BRAIN. Broad Research in Artificial Intelligence and Neuroscience, 12*(4), 509–531. https://doi.org/10.18662/brain/12.4/262

Bae, S., & Wooldridge, R. (2020). Omnichannel customer experience and technological evolvement in retail. *Journal of Textile Science & Fashion Technology, 4*(5). https://doi.org/10.33552/jtsft.2020.04.000596

Balakrishnan, J., Nwoba, A. C., & Nguyen, N. P. (2021). Emerging-market consumers' interactions with banking chatbots. *Telematics and Informatics, 65*, 101711.

Batra, M. M. (2017). Customer experience-an emerging frontier in customer service excellence. In *Competition forum* (Vol. 15, No. 1, pp. 198–207). American Society for Competitiveness.

Batra, M. M. (2019, July). Strengthening customer experience through artificial intelligence: An upcoming trend. In *Competition forum* (Vol. 17, No. 2, pp. 223–231). American Society for Competitiveness.

Bell, J. E., Bradley, R. V., Fugate, B. S., & Hazen, B. T. (2014). Logistics information system evaluation: Assessing external technology integration and supporting organizational learning. *Journal of Business Logistics, 35*(4), 338–358.

Bilgeri, N. (2020). *Artificial intelligence improving CRM, sales and customer experience: An analysis of an international B2B company.* Doctoral dissertation, FH Vorarlberg (Fachhochschule Vorarlberg).

Bisoi, S., Roy, M., & Samal, A. (2020). Impact of artificial intelligence in the hospitality industry. *International Journal of Advanced Science And Technology, 29*(5), 4265–4276.

Biswas, S., Carson, B., Chung, V., Singh, S., & Thomas, R. (2020). *AI-bank of the future: Can banks meet the AI challenge.* McKinsey & Company.

Bradlow, E. T., Gangwar, M., Kopalle, P., & Voleti, S. (2017). The role of big data and predictive analytics in retailing. *Journal of Retailing, 93*(1), 79–95.

Brooksworth, F., Mogaji, E., & Bosah, G. (2023). Fashion marketing in emerging economies: An introduction. In F. Brooksworth, E. Mogaji, & G. Bosah (Eds.), *Fashion marketing in emerging economies volume I* (Palgrave studies of marketing in emerging economies). Palgrave Macmillan. https://doi.org/10.1007/978-3-031-07326-7_1

Brynjolfsson, E., Hu, Y. J., & Rahman, M. S. (2013). Competing in the age of omnichannel retailing. *MIT Sloan Management Review, 50*(2), 34–56.

Cai, Y., & Lo, C. (2020). Omni-channel management in the new retailing era: A systematic review and future research agenda. *International Journal of*

Production Economics, 229, 107729. https://doi.org/10.1016/j.ijpe.2020.107729

Cao, L., & Li, L. (2015). The impact of cross-channel integration on retailers' sales growth. *Journal of Retailing, 91*(2), 198–216.

Chemma, N., El Amine Abdelli, M., Awasthi, A., & Mogaji, E. (2022). Management and information technology in the digital era: Introduction to edited collection on challenges and perspectives. In N. Chemma, M. El Amine Abdelli, A. Awasthi, & E. Mogaji (Eds.), *Management and information technology in the digital era* (Advanced Series in Management) (Vol. 29, pp. 1–6). Emerald Publishing Limited, Bingley. https://doi.org/10.1108/S1877-636120220000029001

Chen, C. F., & Chen, F. S. (2010). Experience quality, perceived value, satisfaction and behavioral intentions for heritage tourists. *Tourism Management, 31*(1), 29–35.

Chi, S., Chang, C., & Chang, C. (2019). We are exploring the application of both internet of things and artificial intelligence under the Omni Channel from the perspective of drama theory. *Proceedings of the 2019 5Th international conference on industrial and business engineering*. https://doi.org/10.1145/3364335.3364343

Chiu, M. C., & Lin, K. Z. (2018). Utilizing text mining and Kansei Engineering to support data-driven design automation at conceptual design stage. *Advanced Engineering Informatics, 38*, 826–839.

Dwivedi, Y. K., Ismagilova, E., Hughes, D. L., Carlson, J., Filieri, R., Jacobson, J., et al. (2021). Setting the future of digital and social media marketing research: Perspectives and research propositions. *International Journal of Information Management, 59*, 102168.

Fagerjord, A., & Kueng, L. (2019). Mapping the core actors and flows in streaming video services: What Netflix can tell us about these new media networks. *Journal of Media Business Studies, 16*(3), 166–181.

Frishammar, J., Cenamor, J., Cavalli-Björkman, H., Hernell, E., & Carlsson, J. (2018). Digital strategies for two-sided markets: A case study of shopping malls. *Decision Support Systems, 108*, 34–44.

Gabisch, J. A., & Gwebu, K. L. (2011). Impact of virtual brand experience on purchase intentions: The role of multi-channel congruence. *Journal of Electronic Commerce Research, 12*(4), 302.

Grewal, D., Motyka, S., & Levy, M. (2018). The evolution and future of retailing and retailing education. *Journal of Marketing Education, 40*(1), 85–93. https://doi.org/10.1177/0273475318755838

Hadida, A. L., Lampel, J., Walls, W. D., & Joshi, A. (2021). Hollywood studio filmmaking in the age of Netflix: A tale of two institutional logics. *Journal of Cultural Economics, 45*, 213–238.

Hallikainen, H., Savimäki, E., & Laukkanen, T. (2020). Fostering B2B sales with customer big data analytics. *Industrial Marketing Management, 86*, 90–98. https://doi.org/10.1016/j.indmarman.2019.12.005

Jain, V., Mogaji, E., Sharma, H., & Babbili, A. S. (2022). A multi-stakeholder perspective of relationship marketing in higher education institutions. *Journal of Marketing for Higher Education*, 1–19. https://doi.org/10.1080/08841241.2022.2034201

Jayakrishnan, S. (2022). Artificial intelligence (AI) in retailing- a systematic review and research agenda. *SSRN Electronic Journal*. https://doi.org/10.2139/ssrn.4134959

Jiang, J., Dong, Y., Li, B., Gu, H., & Yu, L. (2020). Do feelings matter? The effect of leader affective presence on employee proactive customer service performance. *International Journal of Contemporary Hospitality Management, 32*(7), 2305–2323.

Jordan, M. I., & Mitchell, T. M. (2015). Machine learning: Trends, perspectives, and prospects. *Science, 349*(6245), 255–260.

Kaliraj, P., & Devi, T. (2021). *Artificial intelligence theory, models, and applications*. CRC Press.

Karp, P. (2016, 2016, baw 150). Can we replace curation with information extraction software? *Database*. https://doi.org/10.1093/database/baw150

Khalil, A., & Abdelli, M. E. A. (2022). Do digital technologies influence the relationship between the COVID-19 crisis and SMEs' resilience in developing countries? *Journal of Open Innovation: Technology, Market, and Complexity, 8*(2), 100–109.

Kim, J. J., Lee, J. S., & Han, H. (2023). Tangible and intangible hotel in-room amenities in shaping customer experience and the consequences in the with-corona era. *International Journal of Contemporary Hospitality Management, 35*(2), 657–681.

Kohler, E., Mogaji, E., & Erkan, İ. (2023). Save the trip to the store: Sustainable shopping, electronic word of mouth on Instagram and the impact on cosmetic purchase intentions. *Sustainability, 15*(10), 8036–8045.

Koohang, A., Nord, J. H., Ooi, K. B., Tan, G. W. H., Al-Emran, M., Aw, E. C. X., & Wong, L. W. (2023). Shaping the metaverse into reality: A holistic multidisciplinary understanding of opportunities, challenges, and avenues for future investigation. *Journal of Computer Information Systems, 63*(3), 735–765.

Korhonen, T., Selos, E., Laine, T., & Suomala, P. (2021). Exploring the programmability of management accounting work for increasing automation: An interventionist case study. *Accounting, Auditing & Accountability Journal, 34*(2), 253–280.

Kossakowski, R., Szlendak, T., & Antonowicz, D. (2018). Polish ultras in the post-socialist transformation. *Sport in Society, 21*(6), 854–869.

Krishna, A. (2013). *Customer sense: How the 5 senses influence buying behavior*. Springer.

Kumar, S. P., & Saha, S. (2017). Influence of trust and participation in decision making on employee attitudes in Indian public sector undertakings. *Sage Open, 7*(3), 2158244017733030.

Kumar, V., Rajan, B., Venkatesan, R., & Lecinski, J. (2019). Understanding the role of artificial intelligence in personalized engagement marketing. *California Management Review, 61*(4), 135–155. https://doi.org/10.1177/0008125619859317

Kunz, W. H., Heinonen, K., & Lemmink, J. G. (2019). Future service technologies: Is service research on track with business reality? *Journal of Services Marketing, 33*(4), 479–487.

Lemon, K. N., & Verhoef, P. C. (2016). Understanding customer experience throughout the customer journey. *Journal of Marketing, 80*(6), 69–96.

Li, Y., Li, G., Tayi, G., & Cheng, T. (2019). Omni-channel retailing: Do offline retailers benefit from online reviews? *International Journal of Production Economics, 218*, 43–61. https://doi.org/10.1016/j.ijpe.2019.05.002

Liu-Thompkins, Y., Okazaki, S., & Li, H. (2022). Artificial empathy in marketing interactions: Bridging the human-AI gap in affective and social customer experience. *Journal of the Academy of Marketing Science, 50*, 1–21.

Ma, J., Nagatsuma, T., Kim, S., & Hasegawa, M. (2019). A machine-learning-based channel assignment algorithm for IoT. *2019 international conference on artificial intelligence in information and communication* (*ICAIIC*). https://doi.org/10.1109/icaiic.2019.8669028.

McAfee, A., Brynjolfsson, E., Davenport, T. H., Patil, D. J., & Barton, D. (2012). Big data: The management revolution. *Harvard Business Review, 90*(10), 60–68.

McNeal, M. L., & Newyear, D. (2013). Introducing chatbots in libraries. *Library Technology Reports, 49*(8), 5–10.

Mogaji, E., & Nguyen, N. P. (2022a). Managers' understanding of artificial intelligence in relation to marketing financial services: Insights from a cross-country study. *International Journal of Bank Marketing, 40*(6), 1272–1298.

Mogaji, E., & Nguyen, N. P. (2022b). The dark side of mobile money: Perspectives from an emerging economy. *Technological Forecasting and Social Change, 185*, 122045.

Mogaji, E., Soetan, T. O., & Kieu, T. A. (2020a). The implications of artificial intelligence on the digital marketing of financial services to vulnerable customers. *Australasian Marketing Journal, 29*(3), 235–242.

Mogaji, E., Olaleye, S., & Ukpabi, D. (2020b). Using AI to personalise emotionally appealing advertisement. In N. Rana et al. (Eds.), *Digital and social media marketing. Advances in theory and practice of emerging markets*. Springer. https://doi.org/10.1007/978-3-030-24374-6_10

Mogaji, E., Farquhar, J. D., Van Esch, P., Durodié, C., & Perez-Vega, R. (2022). Guest editorial: Artificial intelligence in financial services marketing. *International Journal of Bank Marketing, 40*(6), 1097–1101.

Mogaji, E., Restuccia, M., Lee, Z., & Nguyen, N. P. (2023a). B2B brand positioning in emerging markets: Exploring positioning signals via websites and managerial tensions in top-performing African B2B service brands. *Industrial Marketing Management, 108*, 237–250.

Mogaji, E., Wirtz, J., Belk, R. W., & Dwivedi, Y. K. (2023b). Immersive time (ImT): Conceptualizing time spent in the metaverse. *International Journal of Information Management, 72*, 102659.

Moore, S., Bulmer, S., & Elms, J. (2022). The social significance of AI in retail on customer experience and shopping practices. *Journal of Retailing and Consumer Services, 64*, 102755.

Mullainathan, S., & Spiess, J. (2017). Machine learning: An applied econometric approach. *Journal of Economic Perspectives, 31*(2), 87–106.

Nair, K. V. (2021). Artificial intelligence (AI): Improving customer experience (CX). In *Artificial intelligence theory, models, and applications* (pp. 75–92). Auerbach Publications.

Narayanan, S., & Nandagopal, R. (2016). From Multi-Channel to Omni-Channel: Determinants of channel choice. *Asian Journal of Research in Social Sciences and Humanities, 6*(9), 1016. https://doi.org/10.5958/2249-7315.2016.00851.0

Nasteski, V. (2017). An overview of the supervised machine learning methods. *Horizons, 2*(4), 51–62.

Nguyen, T. M., Quach, S., & Thaichon, P. (2022). The effect of AI quality on customer experience and brand relationship. *Journal of Consumer Behaviour, 21*(3), 481–493.

Nuruzzaman, M., & Hussain, O. (2018). A survey on Chatbot implementation in customer service industry through deep neural networks. *2018 IEEE 15Th international conference on E-business engineering (ICEBE)*. https://doi.org/10.1109/icebe.2018.00019

Pal, S., De, D., & Buyya, R. (2022). *Artificial intelligence-based internet of things systems*. Springer.

Paluch, S., & Wirtz, J. (2020). Artificial intelligence and robots in the service encounter. *Journal of Service Management Research, 18*(4), 261–289.

Panda, S., & Chakravarty, R. (2022). Adapting intelligent information services in libraries: A case of smart AI chatbots. *Library Hi Tech News, 39*(1), 12–15. https://doi.org/10.1108/lhtn-11-2021-0081

Pansari, A., & Kumar, V. (2017). Customer engagement: The construct, antecedents, and consequences. *Journal of the Academy of Marketing Science, 45*(3), 294–311.

Paschen, U., Pitt, C., & Kietzmann, J. (2020). Artificial intelligence: Building blocks and an innovation typology. *Business Horizons, 63*(2), 147–155.

Pelet, J., Lick, E., & Taieb, B. (2019). Internet of things and artificial intelligence in the hotel industry: Which opportunities and threats for sensory marketing? *Advances in National Brand and Private Label Marketing*, 154–164. https://doi.org/10.1007/978-3-030-18911-2_20

Raisch, S., & Krakowski, S. (2020). Artificial intelligence and management: The automation-augmentation paradox. *Academy of Management Review*. https://doi.org/10.5465/2018.0072

Rosário, A. (2021). The background of artificial intelligence applied to marketing. *Academy of Strategic Management Journal, 20*(1), 1–19.

Russell, S. J., & Norvig, P. (2016). *Artificial intelligence: A modern approach*. Malaysia.

Rusthollkarhu, S., Toukola, S., Aarikka-Stenroos, L., & Mahlamäki, T. (2022). Managing B2B customer journeys in digital era: Four management activities with artificial intelligence-empowered tools. *Industrial Marketing Management, 104*, 241–257. https://doi.org/10.1016/j.indmarman.2022.04.014

Salami, O., & Mnkandla, E. (2021). Towards a machine learning enabled multichannel messaging framework for financial service institutions: Preliminary investigations. *2021 International conference on artificial intelligence, big data, computing and data communication systems (Icabcd)*. https://doi.org/10.1109/icabcd51485.2021.9519358.

Salami, O., & Mtsweni, J. (2016, May). Towards a context-aware multi-channel messaging model for African banks: Preliminary investigations. In *2016 IST-Africa week conference* (pp. 1–9). IEEE.

Sattarapu, P. K., Wadera, D., Nguyen, N. P., Kaur, J., & Kaur, S. (2023). Tomeito or Tomahto: Exploring consumer's accent and their engagement with artificially intelligent interactive voice assistants. *Journal of Consumer Behaviour*. https://doi.org/10.1002/cb.2195

Shabbir, J., & Anwer, T. (2018). *Artificial intelligence and its role in near future.* arXiv preprint arXiv:1804.01396.
Shankar, V. (2018). How artificial intelligence (AI) is reshaping retailing. *Journal of Retailing, 94*(4), vi–xi.
Sharma, H., Soetan, T., Farinloye, T., & Noite, M. D. F. (2022). AI adoption in universities in emerging economies: Prospects, challenges and recommendations. In E. Mogaji, V. Jain, F. Maringe, & R. E. Hinson (Eds.), *Re-imagining educational futures in developing countries.* Palgrave Macmillan. https://doi.org/10.1007/978-3-030-88234-1_9
Shen, X. L., Li, Y. J., Sun, Y., & Wang, N. (2018). Channel integration quality, perceived fluency and omnichannel service usage: The moderating roles of internal and external usage experience. *Decision Support Systems, 109*, 61–73.
Snyder, H. (2019). Literature review as a research methodology: An overview and guidelines. *Journal of Business Research, 104*, 333–339.
Steinhoff, L., Arli, D., Weaven, S., & Kozlenkova, I. (2018). Online relationship marketing. *Journal of the Academy of Marketing Science, 47*(3), 369–393. https://doi.org/10.1007/s11747-018-0621-6
Verhoef, P. C. (2021). Omni-channel retailing: Some reflections. *Journal of Strategic Marketing, 29*(7), 608–616.
Von Briel, F. (2018). The future of omnichannel retail: A four-stage Delphi study. *Technological Forecasting and Social Change, 132*, 217–229.
Voorveld, H., Smit, E., Neijens, P., & Bronner, A. (2016). Consumers' cross-channel use in online and offline purchases: An analysis of cross-media and Cross-Channel behaviors between products. *Journal of Advertising Research, 56*(4), 385–400. https://doi.org/10.2501/jar-2016-044
Wilson, H. J., & Daugherty, P. R. (2018). Collaborative intelligence: Humans and AI are joining forces. *Harvard Business Review, 96*(4), 114–123.
Wirtz, J., Patterson, P. G., Kunz, W. H., Gruber, T., Lu, V. N., Paluch, S., & Martins, A. (2018). Brave new world: Service robots in the frontline. *Journal of Service Management, 29*(5), 907–931.
Wynne, K. T., & Lyons, J. B. (2018). An integrative model of autonomous agent teammate-likeness. *Theoretical Issues in Ergonomics Science, 19*(3), 353–374.
Zhang, S., & Zhu, D. (2020). Towards artificial intelligence-enabled 6G: State of the art, challenges, and opportunities. *Computer Networks, 183*, 107556. https://doi.org/10.1016/j.comnet.2020.107556

9

Data Processing and AI-Technology Integration for Personalized Services

Gourav Roy, Varsha Jain, and Parth Salunke

Introduction

Personalization has become one of the popular strategies in the services industry (Gwinner et al., 2005; Shen & Ball, 2009; Sheth et al., 2022), and customers expect personalization. Companies in the service sector need to offer personalization as it helps to improve relationships and increase revenue (Hyken, 2017; Kostusev, 2018). A Forbes report found that out of the 1000 respondents surveyed, a high number of respondents demanded personalized services and wanted more than impersonal experiences (Hyken, 2017). Today, the service sector uses advanced technologies to personalize offerings to its customers. It is about facilitating adjustments and technologically mediated modifications to provide several recommendations to customers based on their individual needs and requirements (Shen & Ball, 2009). Before technological advancements,

G. Roy • V. Jain (✉) • P. Salunke
MICA, Ahmedabad, India
e-mail: gouravroy.fpm20@micamail.in; varsha.jain@micamail.in; parthsalunke.fpm21@micamail.in

© The Author(s), under exclusive license to Springer Nature Switzerland AG 2023

J. N. Sheth et al. (eds.), *Artificial Intelligence in Customer Service*,
https://doi.org/10.1007/978-3-031-33898-4_9

personalization was even offered with human-to-human interactions such as telephonic conversation and face-to-face interactions that enabled the service representative to understand customer needs before service offering. However, with the advent of internet technologies, the personalization of services, such as various recommendations, customer queries, and feedback, is facilitated by artificial intelligence (AI)-based technologies with no or less human involvement. For example, Starbucks, a popular coffee outlet, has been using AI-based technology such as Deep Brew to personalize customer experiences. They offer service-related personalized recommendations to customers for coffee consumption based on weather, past consumption, and many others through a collection of customer data points (Hyperight, 2021). Even the company uses GPS-based tracking through the mobile application to send a personalized offering of services to its customers (Qrius, 2022).

This technology-integrated personalization takes place with the help of different software solutions which utilize the customer data by processing and refurnishing it for AI-based technologies (Grieve, 2022). Even the CX Trend (2022) report found that customers are comfortable sharing their data. Thus, customer data is leveraged by the service industries such as hotels, restaurants, hospitals, and service delivery platforms to offer personalized services. Here, customer data is a key for AI-based technologies to understand customer needs leading to offering the personalization of services. This personalization leads to better experiences, as industry reports claim that 76% of surveyed respondents expect personalized service experiences (Grieve, 2022). For instance, the popular online platform Netflix uses the customer data available to offer personalized recommendations through AI-based technology integration. They facilitate customer experiences by altering the captions and images under every movie recommendation based on the choice and preferences of an individual customer (Basilico, 2019). Thus, based on the received customer data, it goes through processing stages, including AI-based technology integration to offer service personalization.

Recent studies indicate that using relevant technology integration helps service providers ensure better service delivery (Khalil & Abdelli, 2022; Payne et al., 2021), wherein AI-technology integration is the most significant differentiator (Bennett & Robertson, 2021). Thus, AI is considered the primary catalyst in providing personalization of services (Han, 2021). This personalization of services offers customer satisfaction and better experiences. However, it requires adequate information and better implementation with the challenge of providing personalization based on available data. Moreover, conceptual frameworks must explain how data processing and technology integration occur from a marketing standpoint. The process is complex; thus, to simplify, this chapter introduces a framework (Fig. 9.1) to explain how customer data is first collected and passes through data processing stages. Here, AI-technology integration takes place, enabling service industries to offer personalization. The entire process is a cyclical loop and continues to provide personalization to customers based on their retrieved data.

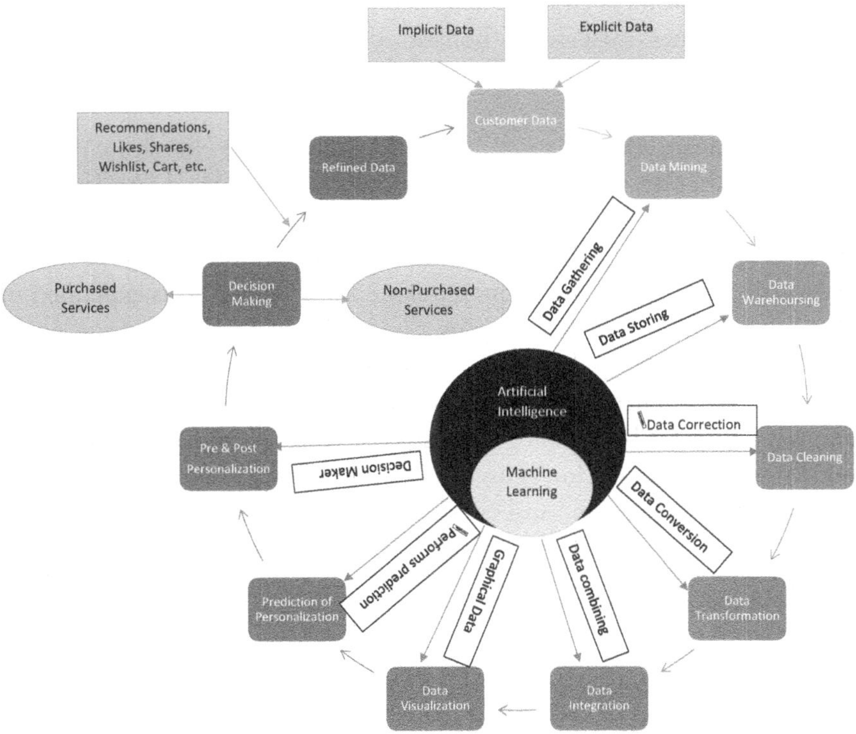

Fig. 9.1 A framework of data processing and AI-technology integration for personalized services

Moreover, the role of AI and personalization appreciated in academic discussions needs to be improved. Earlier, discussions could have been more complex, as theoretical synthesis cited only the relevance and benefits of AI-based personalization (Sahu et al., 2015). Prior studies have indicated AI and personalization tools as intelligent business entities, whereas users' roles have yet to be noticed. Our chapter proposes to minimize the ongoing gaps and profoundly explains the relevance of data processing and technology integration facilitated by AI-centric personalization. Subsequently, the chapter describes the cyclical data processing process and possible AI integration sources adding value to extend the Engel-Kollat-Blackwell (EKB) model of decision-making. Our discussions highlight customers' considerable relevance, even though personalization is pivotal. However, technology and AI centrism are successful only when customers enjoy an elite service personalized experience. The chapter outlines AI-centric holistic data processing and customer decision-making choices in an AI-technology-enabled personalized service environment.

Furthermore, the chapter with the integrative literature review approach discusses the framework's components in detail. Also, it outlines that machine learning is a subset of AI that helps predict personalization, and further, the AI acts as the facilitator of decisions for offering personalization. Figure 9.1 represents the framework and is explained in detail below. The further sections divide the chapter into four parts. Section 'Data Processing Stages and AI-Technology Integration for Personalized Services' details each component of the data processing framework and AI-technology integration that takes place to offer personalized services. Section 'Discussion and Conclusion' presents the discussion and conclusion, and Section 'Contributions and Implications' provides the theoretical contribution and implications to industry and academia.

Data Processing Stages and AI-Technology Integration for Personalized Services

In this section, the components of the framework are explained in detail. It outlines how data processing beings, from collecting the customer data points to undergoing the cyclical loop process for offering AI-technology-based personalization of services. Through the entire framework of data processing, this chapter adds value to the EKB model of decision-making. The model proposed in 1972 enunciates four elements wherein the information is received through marketing stimulus and undergoes information processing leading to decision-making choices affected by external factors such as family and peer group influences (Jain et al., 2019). However, this theoretical framework of Engel-Kollat-Blackwell can be extended further in AI-technology integration for offering personalized services. The chapter proposes that the customers are the value-creation actors in the decision-making process. The AI-technology integration collects customer data points, and customer information is utilized in enhancing the stimulating factors by offering service personalization. Thus, the decision-making process is a cyclical loop wherein customer data is used to provide them with input (stimulating factors), which further undergoes information processing and decision-making choices and is affected by external elements. Additionally, basis the customer's decision for the offered services, the customer data points are again collected, and the entire process repeats through AI-technology integration.

The framework is explained in detail in each section below, with the primary objective of data processing and AI-technology integration. The process starts with collection and customer data points, explained below.

Customer Data

The term 'data' is widely used today and is considered the new oil. Customer data is essential for organizations to offer personalized recommendations for services. Organizations receive customer data from various online channels shared by the customer while using various online sites or applications knowingly or unknowingly. Further, companies use

Table 9.1 Customer data types

Data categories	Implicit data	Explicit data
Definition	The data that the customer shares unintentionally is called implicit data.	The data shared by the customer intentionally is called explicit data.
Data Types	Social media insights, internet browsing history, viewing the services, third-party cookies, etc.	Contact details, email Id, personal information such as weight, gender and responding to detailed surveys, feedback on online sites, etc.
Examples	International hotels such as Radisson Blu collect the implicit data of the customers. This data helps them to track past customers, observe behavior patterns, etc., and enables the personalization of the target audience.	At the primary level, various online applications or websites ask the customer to share essential details. For instance, while registering at Amazon, the customer knowingly filled in their basic information and shared it with the organization.

this data to offer personalization to their customers. Data collection is the first step required to begin the data processing stage. The customer data can be divided into two categories: 1) Explicit data and 2) Implicit data (Najafabadi et al., 2017). Both categories of data are explained in Table 9.1 with examples.

Thus, customer data is classified into two categories. However, explicit data is more valuable to businesses as it offers detailed specifications of the customers, while implicit data helps companies to understand customer behavior and consumption patterns (Clayton, 2019). The implicit data is more difficult as it does not include clarifications; however, with consistent behavioral patterns of customers, it is possible to predict their personal choices and preferences.

Data Mining

This is the second step, where the data processing begins. The collected customer data goes through this step initially. Data gathering is data mining, collecting data to discover new patterns and knowledge from large data sets (Han et al., 2012; Mogaji et al., 2020). It facilitates understanding usage patterns by gathering them through the customer's activity. For instance, when a customer looks for their inquiry into a specific hotel service in a particular location, the data mining algorithm collects the information through search engine metrics (Mariani & Borghi, 2021).

Moreover, banks and financial institutions regularly invest in data mining technologies because it helps them understand their customer's economic mentality. If a system detects that a certain number of users are looking for insurance plans, the service providers recommend services according to the customer's search (Abdulquadri et al., 2021; Taimoor & Rehman, 2021). However, data mining is the initial step in data processing. The collected data at the data mining stage go through other processes to enable personalized service accurately. Subsequently, the data mining process helps a business provider to serve better by utilizing digital gateways (Buhalis & Moldavska, 2021; Mogaji & Nguyen, 2022). Data mining algorithms present patterns and metrics by integrating organizational goals, which have been beneficial for businesses as well as for customers. Various data mining techniques are used by different organizations, such as association rules, classification analysis, clustering, etc., and are used for different purposes based on the service provider's objectives (Han et al., 2012; Kotsiantis, 2006). Overall, data mining presents an in-depth analysis of data patterns and metrics in a service engine.

Data Warehousing

After the data mining stage, the data gathered gets stored at data warehousing. The organization's strategic information, including customer knowledge, is stored here. Data warehousing helps the organization by providing direct access to customers' data and helps analyze business trends and performance (Krishnan, 2013; Ponniah, 2004). In data warehousing, the data is stored after data mining, which helps to create patterns. This stored data at warehousing is essential for the personalization of services (Mingotto et al., 2021) as it helps channel the data sources per the organization's requirement. Here the organization receives data that becomes more accessible, consistent, and of high quality enabling better decision-making (Simon, 2020). The data in warehousing at the organizational level can be of three types, in general, such as (1) Enterprise data warehouse, (2) Data mart, and (3) Operational data store (Tableau, n.d.-b)

1. *Enterprise data warehouse:* This warehouse collects the data or a group of data in a centralized and more organized format and is utilized per the organizational needs (Snowflake, n.d.)
2. *Data mart:* This is also called a data lake which is more centered toward a specific location, meaning used for a business or a particular line of business such as sales. Here a specific team can access the data and receive the required information quickly (Tableau, n.d.-a).
3. *Operational data store:* This type of data storage is a hybrid warehouse that stores data and reports solutions. Here, it collects the data from several sources by consolidating and integrating it for real-time analysis (Buuck, 2022).

Data Cleaning

The data that appears from the real world may include some gaps; it may be noisy or inconsistent. Hence, after data warehousing immediately, data cleaning takes place. Data cleaning is a process that tries to fix the discrepancies in the data, such as handling the missing values found in the data set. At this stage, the data is corrected, and incomplete or incorrect data quality is improved (Chapman, 2005). During the data cleaning stage, the AI and ML functions where the data gets accumulated into a particular model or format. This further helps to provide possible recommended services at a later stage (Krishnan et al., 2016). The data cleaning process facilitates the data to be used by AI algorithms making it clean without any errors and removing all the detected errors in the data. For instance, to provide personalized hospitality services, the data is gathered from various sources such as social media, demographics, contact details, etc. This data collected at warehousing may have errors or duplications removed through data cleaning. Thus, data cleaning is a fundamental and essential data processing step (Tableau, n.d.-b) that helps to remove duplicated, uncorrupted errors and mislabeled data. However, there are no defined steps for data cleaning as the process varies from dataset to dataset. But, for better clarity, a few steps are provided below.

1. *Step 1: Removing duplications and extraneous observations*: Data gets gathered from multiple sources/departments, so there is a high chance that the data gets duplicated. Hence, de-duplication is the primary process conducted in the data cleaning stage. Further, one may receive extraneous observations in the data that may affect the outcome. For instance, a hotel predicts the personalization of services by understanding Generation Z customers. Still, the data may have an observation of some millennials, and hence the output may get distracted. Thus, it is mandatory to remove such irrelevant observations to maintain minimum distraction and an efficient dataset.
2. *Step 2: Fixing the structural level errors*: In the next step, the data may have various anecdotes while you transfer the data, such as different naming patterns, typing errors, and incorrect capitalization. For instance, the data may have inconsistency received while responding to general questions such as 'Not Applicable'; the variations can be N/A, n/a, NA, etc. Here the errors must be fixed by analyzing the responses into the same category.
3. *Step 3: Filtering the outliers*: In the dataset, there will often be observations that do not fit the analysis, such as improper entry or applying filters to your data to remove the outliers. Thus, based on the validity of the data, consider removing the outliner observations if it is mistakes or irrelevant observations. For instance, the dataset considers the luxury consumers determined basis the classification of income group and if the income group of some consumers does not fit for luxury services. Then, these are outliners that may affect the results, and hence it is advisable to remove the outliners.
4. *Step 4: Treating the missing data*: In the dataset, there are chances that some data is missing, and algorithms may not accept the missing values. Thus, handling the missing values effectively is required for better outcomes. There are various ways to deal with these missing values. First, one may consider dropping those observations of a respondent with a missing value. However, this may facilitate the dropping of other relevant information. One should check how vital the observa-

tion is for the analysis. Second, missing values can be added based on other observations in the dataset that leads to an assumption of the data rather than the actual observation of data. Hence, it losses the integrity of the data. Third, the data usage pattern can be altered to navigate the null values.

5. *Step 5: Validating the data*: The last step is validating the entire dataset by answering a few questions at the end of the data cleaning stage. The questions can be related to the following parameters: checking the data based on its objective, following appropriate rules, proving any phenomena or bringing any fruitful insights, and last, if trends in data facilitate new findings. If the answer to this parameter is negative, then a data quality issue needs further cleaning.

Data cleaning is essential, and false or errored data can affect business decisions and strategy. Hence, the data should be of good quality, and various software for data cleaning, such as Tableau prep, can be more effective in forming better data and receiving quality results/outcomes for business decision-making.

Data Transformation

After the data cleaning stage, a massive volume of data is accumulated and further assimilated into various system formats. Data transformation is all about converting data from one format to another. This conversion helps to make better decisions and improvises the outcomes for the personalization of services (Sharma, 2020). Here the data devoid of corruption or errors are transformed into three structures: 1) Structured data, 2) Semi-structured data, and 3) Unstructured data.

1. *Structured data*: This is descriptive data that includes the set of information in a particular fashion per the data presentation model requirements (IBM, 2021). It is the modeled data or formatted data which has a well-defined structure. This data is managed by the Structured Query Language (SQL) database system and helps the machine to understand or interpret the data easily. Moreover, this data is pre-

sented in fixed rows and columns and is received explicitly from the customer or implicitly generated by the machine (Cubeware Gmbh, 2022). For instance, the structured data includes names, phone numbers, credit card details, addresses, postal codes, states, etc.

2. *Semi-structured data*: This format generally does not get into modeled data but includes some structure. Unlike structured data, this semi-structured data cannot be stored in a fixed arrangement of rows and columns (Sint et al., 2009). But, this data format is partially structured and has some definite structures but does not fit into the structured data format (Naeem, 2020). The best example of semi-structured data is delimited files such as emails, zip files, or web pages. This data has structured attributes, but there is a need to provide a structure to this data.
3. *Unstructured data*: This data is in absolute raw format, and it is challenging to include it in a specific structure (Kumar & Khosla, 2018). The data has complex arrangements, making analysis difficult on the unstructured dataset. It can be qualitative or quantitative data such as audio recordings, satellite imagery, pictures, videos, reports, etc. (Naeem, 2020). However, this data can help the service organization to figure out future consumption trends and the effectiveness of their marketing campaigns. This further can be processed through AI-based interpretation to derive meanings from unstructured and semi-structured data. Here, natural language processing (NLP), a part of AI, intervenes to understand better and interpret the data. For instance, Facebook utilizes NLP in its messenger to extract information such as data and time that businesses receive from their texts.

The crucial question is how unstructured or semi-structured data is converted into structured data. Even if there is a need to convert unstructured data to structured data? The unstructured data format can generate more significant insights to help business decision-making. NLP is beneficial in interpreting and understanding unstructured data. However, there are various steps for utilization of unstructured data through conversion that includes 1) cleaning of unstructured data, 2) selection of technology for data collection and warehousing basis the organization's requirements, 3) extracting the internal relationships between the data,

4) categorize the data to create patterns basis the context, 5) analyze the data through machine learning technologies, 6) check the necessary outcomes received through analysis, and 7) store the data (data lakes) for future usage. Various software such as 'Hadoop' store and process large datasets using simpler programming models.

Data Integration

This is one of the essential strategies for service organizations. During this process, the consumed data are combined so the organization can leverage all the categories of data (Lenzerini, 2002). This step may only be necessary for some organizations as it depends on the required data and its usage by organizations with diverse business landscapes (SAP, n.d.). For instance, Hyatt, a hotel chain, has various locations and geographies. Here it becomes necessary for the hotel to combine the data and understand what's right or wrong across the geographical areas. Data integration allows the organization to correct decisions and minimize risk. Thus, data integration brings all the activities of different organizations together to effectively deliver the personalization of services to its customers.

Data Visualization

This step enables the visualization of the data with the help of computer screens. Here the data gets converted into various graphical elements such as charts and plots, helping convey the complex data more meaningfully and thoughtfully (IBM Cloud Education, 2021). However, these visualizations still need to be polished to offer personalization. It helps to only represent the data in a graphical format for a machine or human to undertake better decisions. It prepares a machine learning algorithm for future predictions to offer personalized services to its customers. This step seeks to communicate the information more effectively by demonstrating patterns, trends, and connections between various data items (Blackwood, 2021). There are multiple methods of visualization (IBM Cloud Education, 2021) that are explained briefly below:

1. *Tables:* These include the data in a structured format with rows and columns representing the usage of comparable data variables such as advertisement clicks and sales revenue generation.
2. *Line charts and area charts:* With predictive analytics, they represent the series of data points used frequently, which includes the presentation of the change in one or more quantities of information in a data set.
3. *Pie charts and Bar charts*: Here, the data representation includes a straightforward structure that presents certain parts of the data set and provides the facility to organize the data more simply and understandably.
4. *Scatter plots*: These types of graphical plots are used when the data set represents the relationship between the two variables. For example, the consumption intent of hotel services for a holiday destination can be positively related customer's previous experience.
5. *Treemaps:* This represents the hierarchical data and is very useful to compare the data of various proportions of various categories or area sizes. For instance, the service effectiveness of various outlets of restaurants can be known through the data set representing different geographical locations.
6. *Heat maps:* They are effective in representing behavioral data based on locations. For example, Starbucks tracks the global positioning system (GPS) location of the customers to offer quick and personalized services. This can be done with the help of heat maps.

Prediction of Personalization: Role of Machine Learning

The data received after data visualization is refined and further used to predict customers' needs and requirements. The role of machine learning is essential at this stage. Machine learning is a subset of artificial intelligence. In their book, Mohri et al. (2018) define a machine learning algorithm as a computational method to offer accurate predictions and better personalization performance. For example, Apple, a phone brand, utilizes machine learning technology to predict the personalization of its customer services.

Another instance is the 'MakeMyTrip' website, which helps to provide booking options based on customer data. Here, the prediction for the results first emerges from machine learning algorithms before it reaches the personalization of services to the customer screen. This prediction is the facilitator to provide personal experiences to the customers that are different for different individuals based on their preferred choices and requirements. Another example is Booking.com, where a property can be rented. Assume that out of 5 individuals visiting a particular property are asked to give some rating. Based on this rating, if two individuals have rated similarly, the machine learning algorithm will suggest or predict similar properties for those two individuals based on their previous rating behavior. Thus, machine learning makes predictions more based on statistical parameters. Further, machine learning can be divided primarily into three types: 1) Supervised learning and 2) Unsupervised learning, and 3) Reinforcement learning. Each of these types is explained in brief below.

1. *Supervised learning*: This is mainly known as labeled data such as name, number, etc., with guided learning models that classify the data to make accurate predictions.
2. *Unsupervised learning*: This is mainly known as unlabeled data, in which the algorithms must identify patterns in the data set.
3. *Reinforcement learning*: This is primarily known as a behavioral machine learning model similar to supervised learning, but here the model runs based on the trial-and-error method.

Further, different machine learning models enable better predictions and offer personalized services. Some overlap with the operating parameter; however, they are explained briefly below (Pedamkar, n.d.)

1. *Regression*: This supervised learning model analyzes regression to predict the numeric value. For example, in a restaurant setting, it can help to expect service consumption during the festive season based on available data.
2. *Classification*: This supervised learning model categorizes the data into groups. For instance, based on service consumption at the hospital, it

classifies the data into satisfied or dissatisfied customers and then helps predict the personalization of services.

3. *Association rule*: This is an unsupervised learning model that infers a certain amount of association between two data points of a customer. This association will be further used for the prediction of personalization. For instance, if customers purchase a health checkup service at home, they might be more likely to approach the doctor online. This prediction is based on previous and similar customer patterns, age, behavioral patterns, etc.
4. *Clustering*: This can be used in unsupervised and supervised learning. The clustering model groups similar things together and even helps detect customer data irregularities that are further used for prediction. This is very helpful in customer segmentation or identifying behavioral patterns. For example, over-the-top (OTT) platforms predict and even offer a similar genre of movies or series to customers based on their choices and preferences.
5. *Neural Networks*: This is not linear statistical modeling, but it facilitates the prediction of personalization through deep learning, the data equivalent to the human brain. This can be used in prediction areas with all other models that undertake video, image, and audio recognition. It helps in solving a complex business case. For example, any hotel with franchises worldwide can undertake neural networks to predict the chances of service offering failures in different markets. The results can solve service failures, and personalization can be offered to customer preferences.

Personalization of Services: Role of Artificial Intelligence as Decision-Maker

While machine learning helps in prediction, artificial intelligence facilitates the decision of personalization of services as they think and understand as humans. Artificial intelligence is incorporated into the system enabling decision-making by an optimal solution to offer personalization of services. For instance, AI is widely used in the healthcare industry in multiple ways. One such example is the usage of chatbots by companies such as Ada Health, Florence, etc. They have AI-enabled chatbots that provide customers with personalized care services and reduce waiting time. Thus, once machine learning completes the prediction of possible options, the artificial intelligence in a system performs the tasks of humans, such as thinking, reasoning, learning from past experiences, and, most importantly, making their own decisions (David, 2021). Even McCarthy (1988) defines AI as the 'science and engineering of making intelligent machines.' However, AI has yet to reach a point that can interact and make decisions for people at an emotional level. Here, AI resolves the tasks required by humans and is a self-reliant machine that can offer personalized services to customers. Another example of an AI application is from the world's largest and fastest-growing hospitality industry. The world-famous hotel 'Hilton' uses the AI-enabled robot—'Connie' at the front desk, which offers personalized suggestions to its customers for visiting tourist sites and attractions (Ruel & Njoku, 2020).

Moreover, the application of ML (a subset of AI) and AI-related technologies is growing in banking services. These institutions continuously improve customer experiences based on client understanding, behavior, and prior interactions. For example, 'Erica' is a virtual assistant created by Bank of America as an AI chatbot, taking care of customers' regular activities.

Thus, AI being the decision-maker finds the optimal solution based on the prediction of customer needs and requirements and then provides the output result, which leads to the personalization of services. Here, it is essential to note that personalization can be pre- and post-personalization of services. For instance, if the customer selects a personalized healthcare service and the same customer dislikes or provides an impersonal experience review. In this case, the customer receives the post-personalization of services from the same healthcare provider or competitor.

Another live example is Amazon; if an individual purchases a service delivery of a mobile device on the first visit. Further, when the individual opens the application, they may receive personalized recommendations of other service delivery options aligned to the prior purchase, such as phone covers and protective mobile glasses. Thus, the personalization of services by artificial intelligence can occur depending on customer behavior and the collected data points.

Customer Decision-Making

After the customer receives the personalized recommendation of services, one dwells on decision-making for selecting the final consumption of services. Here, artificial intelligence cannot take place. It is the individual process undertaken by the customer. This decision-making is also a complex process, as even the personalization of services offers several recommendations based on customer data points (Kumar, 2021). However, the AI-enabled recommendation makes customers feel that the service organization understands their needs and requirements. Even Shen and Ball (2009) found that these personalized recommended services are attractive additions leading to efficient consumption. Here, one might think that AI has replaced decision-making, but the customer is the decision-maker given final consumption (Duan et al., 2019). The AI-enabled technology integration is just a facilitator in offering optimal solutions. Thus, based on the provided solutions of services based on a prediction of choices and preferences, the customer assesses the given set of alternatives for service consumption and then makes the selection (Jain et al., 2019). Further, during assessing alternatives performed by customers, the AI filters various options such as price, specific requirements, time, etc. (Zanker et al., 2019).

With the set of alternatives, the customers have two choices: to purchase the service or not purchase a service. This customer decision again generates valuable data in refined form and helps an organization better personalize. Here, the not-purchase services may include the customer's intention to purchase, add to a wishlist, hold or cart, etc. Even if the services are purchased, the customer data is generated through their feedback. Thus, the customer leaves the data for further improvements and personalization in any possible case. It means the personalization of services is a cyclical process that keeps improving based on customer data points.

Thus, according to the EKB model, the stimulus such as advertisements, communication from a customer service agent, service staff, and customer interactions is all generated by AI in various ways such as chatbots and robot-enabled services. Here, the entire decision-making model should be developed as a cyclical loop wherein the customer and its data points are value creators for the stimulating factors in offering service personalization. Further, the entire choice-making process of a model may remain the same. Still, the customer is always a decision-maker and generates more data points for the business organizations through consuming services or not consuming services. The objective of this chapter was to explore the data processing and AI-technology integration parameters that help in consumer decision-making. Further research can redevelop the EKB model more robust through exploratory and confirmatory studies.

Discussion and Conclusion

Data processing and AI-technology integration offer personalization in customer services. Every individual customer receives personal experiences based on their choices and preferences. Even the chapter illustrated how the personalization of services has become essential in customer services, and customers expect personalized experiences. The services sector should leverage the opportunity of AI-aided technologies to offer personalization to its customers. Furthermore, the chapter presents a holistic view with a framework that the entire process of providing personalized services is a cyclical loop. It begins with the customer data (explicit and implicit data). Then this data enters the data processing stages, wherein the role of artificial intelligence is crucial to offer the optimal solutions based on customer needs and requirements. Once the customer receives the personalized service through artificial intelligence, they undergo a decision-making process from the available alternatives. Here, one should note that AI is the decision facilitator for offering certain personalized services; however, the customer makes the final decision on purchasing or not purchasing a service.

Further, based on the selected alternative by a customer, the data gets refined, and the customer receives the post-personalization outcomes. Thus, the process is never-ending and is a cyclical loop. The data processing and AI-technology integration have offered services ease, real-time monitoring, and constant learning of customer choices and preferences. It efficiently allows the personification of options and preferences and is of paramount importance (Ge et al., 2022), wherein the AI acts as a catalyst. Lastly, this chapter aims to elucidate the importance of data processing and AI-technology integration. It also recommends that managers understand the importance of personalization and include pattern maps and unique metrics while aligning customer service strategies. Managers should optimize technology and integrate it for effective service personalization.

Contributions and Implications

The chapter highlights a few theoretical contributions and practical implications. First, this chapter presents the holistic framework of how data processing and AI-technology integration enable services' personalization. Second, the chapter proposes extending the work of Engel-Kollat-Blackwell's decision-making model. It enunciates that customers allow marketers to stimulate factors by collecting customer data points from various sources. Scholars and practitioners should consider the cyclical decision-making model wherein AI-technology integration plays a pivotal role. Third, it extends the ongoing literature by presenting the marketing standpoint and customer decision-making. Fourth, the chapter discussed integrating technology to offer better personalized services, a gateway for academicians and practitioners to develop strategies to provide better personalization. Fifth, the managers should work to offer a unique service experience, and scholars should develop a strategic framework of how personalization can facilitate a unique service experience through AI-technology integration. Sixth, it stresses that managers must give importance to the users during a service transition, and technology integration must facilitate the customer's unique choices, needs, and preferences. Seventh, it proposes that AI-technology integration and data processing is a complex process that can assist the business in offering several recommendations as personalized services. However, the consumer is the final decision-making authority, and the AI may interpret the data wrong, offering an impersonal service experience and may lead to a loss of customer interest in service consumption. Thus, the business should utilize the human and technology to offer personalized services to its customers. Eight, managers should understand the importance of AI-based technology and should leverage it as the opportunity to facilitate better, concrete, and personal experiences in service consumption.

References

Abdulquadri, A., Kieu, T. A., & Nguyen, N. P. (2021). Digital transformation in financial services provision: Perspective to the adoption of chatbot. *Journal of Enterprising Communities: People and Places in the Global Economy, 15*(2), 258–281.

Basilico, J. (2019). *Netflix research.* Retrieved from https://research.netflix.com/research-area/machine-learning

Bennett, K. W., & Robertson, J. (2021). Remote sensing: Leveraging cloud IoT and AI/ML services. In T. Pham & L. Solomon (Eds.), *Artificial intelligence and machine learning for multi-domain operations applications III* (Vol. 11746, pp. 677–691). SPIE.

Blackwood, A. (2021). *Data visualisation: Basics & trends.* Retrieved from https://blog.iotechnologies.com/data-visualization-trends/

Buhalis, D., & Moldavska, I. (2021). In-room voice-based AI digital assistants transforming on-site hotel services and guests' experiences. *Information and Communication Technologies in Tourism 2021, 23*(2), 30–44. Springer, Cham.

Buuck, B. (2022). *The operational data store (ODS) vs data warehouse.* Retrieved from https://streamsets.com/blog/operational-data-stores-ods-data-warehouses/

Chapman, A. D. (2005). *Principles and methods of data cleaning: Primary species and species-occurrence data.* Global Biodiversity Information Facility.

Clayton, T. (2019). *The difference between implicit and explicit data for business.* Retrieved from https://blog.mirumee.com/the-difference-between-implicit-and-explicit-data-for-business-351f70ff3fbf

Cubeware Gmbh. (2022). *Structured data: Definition, examples, and comparison.* Retrieved from https://www.linkedin.com/pulse/structured-data-definition-examples-comparison-cubeware-gmbh/

CX Trend. (2022). *The Zendesk customer experience trends report 2022.* Retrieved from https://www.zendesk.com/customer-experience-trends/

David, D. (2021). *AI vs ML – What's the difference between artificial intelligence and machine learning?* Retrieved from https://www.freecodecamp.org/news/ai-vs-ml-whats-the-difference/

Duan, Y., Edwards, J. S., & Dwivedi, Y. K. (2019). Artificial intelligence for decision making in the era of big data–evolution, challenges and research agenda. *International Journal of Information Management, 48*, 63–71.

Ge, B., Wang, Q., & Yao, M. (2022). From ideas to entrepreneurial opportunity: A study on AI. *Systems Research and Behavioral Science, 39*(3), 618–632.

Grieve, P. (2022). *The importance of providing personalized service in 2022*. Retrieved from https://www.zendesk.com/in/blog/start-providing-personalized-customer-service/#georedirect

Gwinner, K. P., Bitner, M. J., Brown, S. W., & Kumar, A. (2005). Service customization through employee adaptiveness. *Journal of Service Research, 8*(2), 131–148.

Han, J., Pei, J., & Kamber, M. (2012). *Data mining: Concepts and techniques*. Elsevier.

Han, M. (2021). Analysis of the public's intention to use the government's artificial intelligence (AI)-based services: Focusing on public values and extended technology acceptance model. *The Journal of the Korea Contents Association, 21*(8), 388–402.

Hyken, S. (2017). *Personalized customer experience increases revenue and loyalty*. Retrieved from https://www.forbes.com/sites/shephyken/2017/10/29/personalized-customer-experience-increases-revenue-and-loyalty/?sh=62f148054bd6

Hyperight. (2021). Deep brew: Transforming Starbucks into an AI & data-driven company. *Hyperight.com*. https://hyperight.com/deep-brew-transforming-starbucks-into-a-data-driven-company/

IBM. (2021). *Structured vs. unstructured data: What's the difference?* Retrieved from https://www.ibm.com/cloud/blog/structured-vs-unstructured-data

IBM Cloud Education. (2021). *What is data visualization?* Retrieved from https://www.ibm.com/cloud/learn/data-visualization

Jain, V., Sheth, J., Schultz, D., & E. (2019). *Consumer behavior: A digital native, 1e*. Pearson Education India.

Khalil, A., & Abdelli, M. E. A. (2022). Do digital technologies influence the relationship between the COVID-19 crisis and SMEs' resilience in developing countries? *Journal of Open Innovation: Technology, Market, and Complexity, 8*(2), 100–109.

Kostusev, D. (2018). *Personalization in sales: Why it matters and how to achieve it*. Retrieved from https://www.forbes.com/sites/forbesbusinessdevelopmentcouncil/2018/06/11/personalization-in-sales-why-it-matters-and-how-to-achieve-it/?sh=5a0ffd819127

Kotsiantis, S. (2006). *Association rules mining: A recent overview*. Retrieved from https://citeseerx.ist.psu.edu/viewdoc/download?doi=10.1.1.103.6295&rep=rep1&type=pdf

Krishnan, K. (2013). *Data warehousing in the age of big data*. Newnes.

Krishnan, S., Franklin, M. J., Goldberg, K., Wang, J., & Wu, E. (2016). Active clean: An interactive data cleaning framework for modern machine learning. In *Proceedings of the 2016 international conference on management of data*. https://doi.org/10.1145/2882903.2899409.

Kumar, V., & Khosla, C. (2018). Data Cleaning-A thorough analysis and survey on unstructured data. In *2018 8th international conference on cloud computing, data science & engineering* (Confluence). https://doi.org/10.1109/confluence.2018.8442950.

Kumar, V. (2021). *Intelligent marketing: Employing new-age technologies*. SAGE Publishing India.

Lenzerini, M. (2002). Data integration. *Proceedings of the twenty-first ACM SIGMOD-SIGACT-SIGART symposium on principles of database systems - PODS '02*. https://doi.org/10.1145/543613.543644.

Mariani, M., & Borghi, M. (2021). Customers' evaluation of mechanical artificial intelligence in hospitality services: A study using online reviews analytics. *International Journal of Contemporary Hospitality Management., 12*(1), 23–69.

McCarthy, J. (1988). Mathematical logic in artificial intelligence. *Daedalus*, 297–311.

Mingotto, E., Montaguti, F., & Tamma, M. (2021). Challenges in re-designing operations and jobs to embody AI and robotics in services. Findings from a case in the hospitality industry. *Electronic Markets, 31*(3), 493–510.

Mogaji, E., & Nguyen, P. N. (2022). Managers' understanding of artificial intelligence in relation to marketing financial services: Insights from a cross-country study. *International Journal of Bank Marketing, 40*(6), 1272–1298.

Mogaji, E., Soetan, T., & Kieu, T. (2020). The implications of artificial intelligence on the digital marketing of financial services to vulnerable customers. *Australasian Marketing Journal., 29*(3), 235–242.

Mohri, M., Rostamizadeh, A., & Talwalkar, A. (2018). *Foundations of machine learning* (2nd ed.). MIT Press.

Naeem, T. (2020). *Understanding structured, semi-structured, and unstructured data*. Retrieved from https://www.astera.com/type/blog/structured-semi-structured-and-unstructured-data/

Najafabadi, M. K., Mahrin, M. N. R., Chuprat, S., & Sarkan, H. M. (2017). Improving the accuracy of collaborative filtering recommendations using clustering and association rules mining on implicit data. *Computers in Human Behavior, 67*, 113–128.

Payne, E. H. M., Dahl, A. J., & Peltier, J. (2021). Digital servitization value co-creation framework for AI services: A research agenda for digital transformation in financial service ecosystems. *Journal of Research in Interactive Marketing., 56*(1), 236–369.

Pedamkar, P. (n.d.). *Machine learning models*. Retrieved from https://www.educba.com/machine-learning-models/

Ponniah, P. (2004). *Data warehousing fundamentals: A comprehensive guide for IT professionals*. John Wiley & Sons.

Qrius. (2022). *The benefits and the technology behind it: Personalization trend*. Retrieved from https://qrius.com/understanding-the-personalization-trend-the-benefits-and-the-technology-behind-it/

Ruel, H., & Njoku, E. (2020). AI redefining the hospitality industry. *Journal of Tourism Futures, 7*(1), 53–66. https://doi.org/10.1108/jtf-03-2020-0032

Sahu, V., Grover, S., Tulachan, B., Sharma, M., Srivastava, G., Roy, M., et al. (2015). Heavily nitrogen doped, graphene supercapacitor from silk cocoon. *Electrochimica Acta, 160*, 244–253.

SAP. (n.d.). What is data integration? | *SAP insights*. Retrieved from https://www.sap.com/insights/what-is-data-integration.html

Sharma, R. (2020, September 14). *6 methods of data transformation in data mining*. Retrieved from https://www.upgrad.com/blog/methods-of-data-transformation-in-data-mining/

Shen, A., & Ball, A. D. (2009). Is personalisation of services always a good thing? Exploring the role of technology-mediated personalisation (TMP) in service relationships. *Journal of Services Marketing, 23*, 79.

Sheth, J. N., Jain, V., Roy, G., & Chakraborty, A. (2022). AI-driven banking services: The next frontier for a personalised experience in the emerging market. *International Journal of Bank*. Marketing, *40*, 1248.

Simon, A. (2020, June 3). *What is data warehousing? Understand the importance of data structures and architecture*. Retrieved from https://blog.udemy.com/what-is-data-warehousing-understand-the-importance-of-data-structures-and-architecture/#:~:text=Data%20warehousing%20typically%20includes%20only%20structured%20data%20such,many%20different%20sources%20and%20applications%20to%20support%20analytics.

Sint, R., Schaffert, S., Stroka, S., & Ferstl, R. (2009). Combining unstructured, fully structured and semi-structured information in semantic wikis. In *CEUR workshop proceedings* (Vol. 464, No. 12, pp. 73–87). Heraklion.

Snowflake. (n.d.). *What is an enterprise data warehouse?* Retrieved from https://www.snowflake.com/guides/what-enterprise-data-warehouse

Tableau. (n.d.-a). *Guide to data cleaning: Definition, benefits, components, and how to clean your data.* Retrieved from https://www.tableau.com/learn/articles/what-is-data-cleaning

Tableau. (n.d.-b). *What is enterprise data warehousing?* Retrieved from https://www.tableau.com/learn/articles/data-warehousing

Taimoor, N., & Rehman, S. (2021). Reliable and resilient AI and IoT-based personalised healthcare services: A survey. *IEEE Access, 10*, 535–563.

Zanker, M., Rook, L., & Jannach, D. (2019). Measuring the impact of online personalisation: Past, present and future. *International Journal of Human-Computer Studies, 131*, 160–168.

Theme IV

Customer and Employee Wellbeing

10

Use of Artificial Intelligence with Ethics and Privacy for Personalized Customer Services

Damini Goyal Gupta and Varsha Jain

Introduction

This chapter aims to understand the role of using artificial intelligence (AI) with ethics and privacy to provide personalized customer services. The world of customer services changed dramatically post the Covid-19 pandemic (DeMaio, 2020). The new way of serving customers involves meeting their heightened expectations (Sheth, 2020). According to the latest report from PwC, customer experience has even overtaken price, whereby 86% of customers reported paying a higher price to receive a better customer experience (Gladly, 2018; Puthiyamadam & Reyes, 2021). This customer experience consists of proactive, friendly, convenient, and consistent services (MIT Technology Review Insights, 2020; Puthiyamadam & Reyes, 2021) that help customers quickly find an apt solution for their problems. This expectation of customers is met through the use of Artificial Intelligence (AI) in customer support services (Forbes,

D. G. Gupta • V. Jain (✉)
MICA, Ahmedabad, India
e-mail: daminigupta.fpm20@micamail.in; varsha.jain@micamail.in

© The Author(s), under exclusive license to Springer Nature Switzerland AG 2023

J. N. Sheth et al. (eds.), *Artificial Intelligence in Customer Service*,
https://doi.org/10.1007/978-3-031-33898-4_10

2021; Kleinings, 2022; Schneider, 2017). For instance, Zipline, a drone-based delivery services brand based out of San Francisco, USA, uses AI-enabled drones to deliver vital medicines, blood, vaccines, and other personal protective medical equipment to customers in several countries. These AI-enabled drones help Zipline overcome the challenges faced by other related brands in meeting emergency medical requirements due to the unavailability of the delivery person, traffic, accidents, and so on. Hence, Zipline's AI-enabled drones create a record of flying over five million miles and completing more than 1.10 lakh deliveries in the shortest possible time (Violino, 2021). Therefore, Zipline presents an excellent example of AI's capabilities in enabling customer support services to enhance customer experience by providing proactive, convenient, and consistent services effectively and efficiently. Therefore, AI is revolutionizing the customer support services industry (Roldós, 2020). The latest report from Gartner reports a 270% increase in the use of AI in customer support services (Costello, 2019). However, implementing AI in customer support services must minimize the risk of a breach of customer privacy while maintaining ethical boundaries (Ameen et al., 2021). Therefore, AI-enabled customer support services ensure that customers' demands are met quickly while using AI with ethics and privacy to enhance the overall customer experience (Davenport et al., 2020). However, since customers demand personalized experiences, AI-enabled customer support services no longer accommodate a one-size-fits-all model to solve individual customers' problems (Krayewski, 2022). For instance, Singapore Airlines, the flag and luxury airline carrier in Singapore, uses an AI-enabled platform to connect with customers. This platform treats each customer differently despite the device used to communicate with the airlines (Writer, 2019). Hence, letting the customer support services know the instant requirement of the customer based on the type of interaction with the AI-enabled platform. The on-ground staff of the airlines then uses this information to work toward addressing the customer issue quickly. Hence, this leads to an enhanced and personalized customer experience (Rowan, 2019). Therefore, the new-age customer support services aim to create an ecosystem whereby the ethical use of technology helps protect customers' privacy by providing a personalized customer experience (Mogaji et al., 2021; Trivedi, 2022). In this regard, this chapter addresses the need to study the use of ethical AI in customer support services, privacy, and customer experience (Davenport et al., 2020). As previously, the studies have focused on the use of AI (Bock et al., 2020; Chatterjee et al., 2022; Dwivedi et al., 2021; Huang & Rust, 2018; Zhu et al., 2022), ethics and privacy (Ameen et al., 2021; Hagendorff, 2020; Mogaji et al., 2021; Mökander & Floridi, 2021; Morley et al., 2021), and customer experience (Ameen et al., 2021; Kautish & Khare, 2022; Sheth et al., 2020) in customer support services independently. This chapter, thus, attempts to comprehensively study and develop a model for understanding the use of AI with ethics and privacy to provide a personalized experience to customers through customer support services.

The rest of the chapter is structured as follows: Section "Background and Literature Review" understands the evolution of AI in customer support services. Section "Use of AI with Ethics and Privacy for Personalized Customer Services Model" discusses various elements and sub-elements of the model developed to understand the use of AI with ethics and privacy for providing personalized customer services. Section "Functional AI" lists the implications for academia and industry. Finally, Section "Ethical AI" discusses the overall findings of the chapter and presents ideas for further research in the area.

Background and Literature Review

Customer support service is considered crucial to service providers across the globe (Spencer-Matthews & Lawley, 2006). The importance of customer support service was first identified in marketing literature in the 1960s (Heskett, 1962; Simon, 1965). However, almost 60 years later, customer support service is still considered the backbone of service-oriented firms (Sheth et al., 2020). The growing importance of the experience economy (Pine & Gilmore, 1998) and Service-Dominant (SD) logic (Lusch & Vargo, 2006, 2011) has enhanced the importance of customer support service manifold. As a result, customer support service is considered a foundation for improving the customer experience and developing solid customer-brand relationships (Sheth, 2011; Sheth et al.,

2020). The total customer experience across several interactions (Diener, 1984; Lee et al., 2002) leads to the formation of customer perception of a firm. This perception is based on fulfilling customers' psychological and self-fulfillment needs (Spencer-Matthews & Lawley, 2006). Hence, making customer support services the primary source of providing satisfaction and ultimately enhancing the customer experience.

Furthermore, using the latest technology, such as Artificial Intelligence (AI), machine learning, robotics, and so on, compliments customer support services' importance. This use of technology with marketing makes customer support services interdisciplinary, having basics of marketing and advancements in Information Technology (IT) (Dwivedi et al., 2021; Ertel, 2018). Hence, it becomes crucial to study the influence of the latest technology, especially AI, in transforming customer support services.

AI is defined as the simulation of human intelligence in technology that helps in providing value via flexible adaptation facilitated by learning, unlearning, and relearning (Bock et al., 2020; Velazquez, 2022) (Table 10.1). Hence, AI is an advanced technology capable of scrolling through customer data and drawing insights. The unique feature of AI involves its ability to learn from the process and grow its intelligence. Hence, AI is an enabler that compliments human intelligence to decode structured and unstructured customer data to derive insights (Mogaji et al., 2021). Customer support services use AI-enabled chatbots, interactive voice responses (IVRs), robots, and so on to collect and disseminate

Table 10.1 Key terms definition

S. No.	Key Terms	Definition
1	Artificial Intelligence (AI)	AI is defined as the simulation of human intelligence in technology that helps in providing value via flexible adaptation facilitated by learning, unlearning, and re-learning (Bock et al., 2020; Velazquez, 2022).
2	Ethical AI	Ethical AI uses automated intelligent systems that cause no harm to human beings' moral, social, or personal values (Dignum, 2018; Hagendorff, 2020; Siau & Wang, 2020). In other words, ethical AI can be understood as a technological advancement that is transparent, inclusive, and accountable (Mökander & Floridi, 2021; Morley et al., 2021).

customer data (Haptik, 2021). The disseminated customer data help the customer support service providers understand the customer's problems and sentiments. This understanding of sentiments and problems helps enhance the overall customer experience. Hence, AI facilitates customer service providers in building long-term connections with customers. Despite the advantages of using AI in improving the overall customer experience, there is a specific threat to using ethical AI while maintaining customers' privacy.

As AI becomes an integral part of customer support services, the concern related to ethical AI becomes an essential subject of study (Dignum, 2018). Ethical AI is automated intelligent systems that cause no harm to human beings' moral, social, or personal values (Dignum, 2018; Hagendorff, 2020; Siau & Wang, 2020). In other words, ethical AI can be understood as a technological advancement that is transparent, inclusive, and accountable (Mökander & Floridi, 2021; Morley et al., 2021). However, the nascent stage of AI development and use makes it difficult for academicians to provide a comprehensive definition of ethical AI. In this regard, the Monetary Authority of Singapore (MAS), along with tech giants Microsoft and Amazon Web Services, unveiled the world's first principle—FEAT (Fairness, Ethics, Accountability, and Transparency) Principle for developing ethical AI in November 2018 (Siau & Wang, 2020). This principle is in its infancy, and other tech giants, such as Google, IBM, and Microsoft, are working on further enhancing the FEAT principle. Therefore, more research is needed in this area of developing models and frameworks for developing ethical AI.

Similarly, the increased use of AI in customer support services requires many customer data. This requirement of AI to collect, store, and process tons of customer data threatens customer privacy (Bélanger & Crossler, 2011). This threat to customer privacy becomes more severe as most AI processes are hidden (Kaplan & Haenlein, 2019), and customers are unaware of how their data is being used. Hence, customers are skeptical about sharing their data with AI-enabled chatbots, robots, and IVRs. For instance, 82% of the customers in the US are concerned about their privacy while sharing personal data with chatbots or IVRs. However, 64% agreed on feeling safe and secure in receiving a human touch (Kim et al., 2022). Hence, customers prefer human interaction and empathy to share private data (Puntoni et al., 2021). Thus, there exists an urgent need for contemporary research to establish a relationship between the use of AI with ethics and privacy to enhance the overall customer experience. We have developed a model discussed in the following section to understand the role of using AI with ethics and privacy to provide personalized customer services.

Use of AI with Ethics and Privacy for Personalized Customer Services Model

Using AI in customer support enables proactive, friendly, convenient, and consistent customer service. However, using AI in customer support services requires massive customer data. When fulfilled by following the ethical ways of collecting, storing, and analyzing customer data, this requirement of extracting customer data helps customer support services gain customers' trust. Furthermore, keeping customer data safe while maintaining privacy helps build a solid relationship between the brand and the customer. Hence, when customer support services use AI within the boundary of ethics and confidentiality, these services lead to an enhanced customer experience (Mande, 2022; Onorato, 2022). However, the absence of any model/process explaining the relationship between AI, ethics, privacy, and customer experience leads to the development of the model presented in Fig. 10.1. AI with ethics and privacy for personalized customer services (Fig. 10.1) is a model developed to offer a comprehensive pathway to help customers and customer service providers understand the relationship between different elements and sub-elements responsible for using AI to enhance the customer experience. Subsections given next will explain the several components of the model in detail.

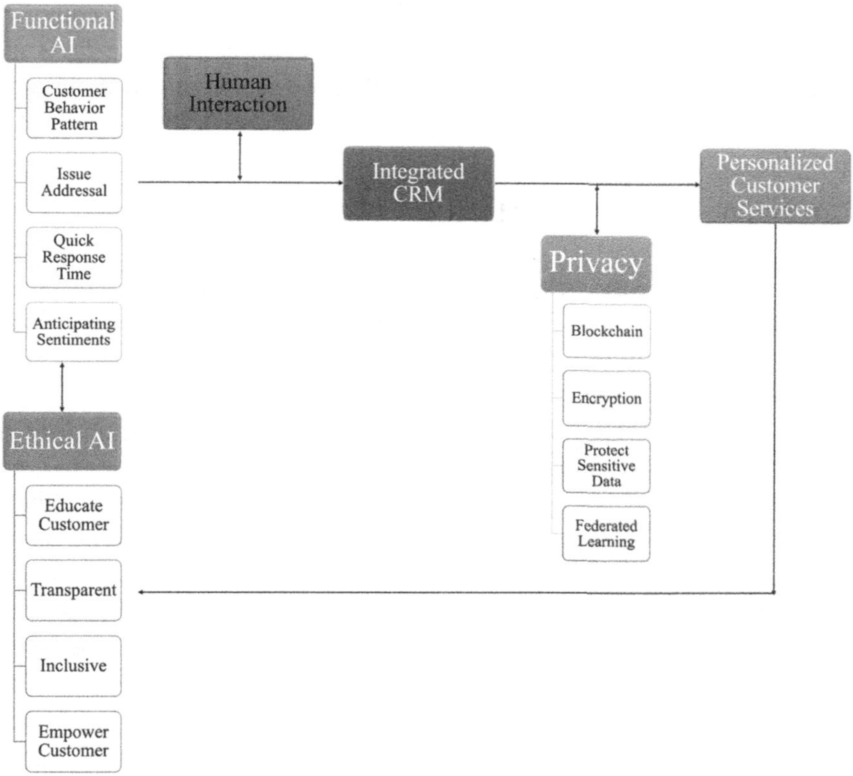

Fig. 10.1 Relationship between AI, ethics, privacy, and customer experience

Functional AI

Functional AI is essential to implementing AI in customer support services. Functional AI helps customer support services automate several functional tasks by processing large amounts of customer data and understanding viral patterns (Davenport & Ronanki, 2018). This capability of AI to parse through colossal customer data to understand customer behavioral patterns and provide a quick address to the issue at the right moment is unmatched.

Furthermore, AI-enabled algorithms' ability to understand customers' sentiments and provide relevant solutions (MIT Technology Review

Insights, 2020) is an added advantage. Hence, functional AI contains several elements that help customer support services enhance the customer experience. The following sub-sections explain the aspects of functional AI in detail.

Customer Behavior Pattern

The first and foremost element of functional AI is analyzing customer behavior patterns. The powerful and trainable AI-enabled algorithms analyze the enormous amount of customer data and learn patterns to provide automation and assistance to customer service processes (Doll, 2019). The AI-enabled algorithms investigate customer data and learn from the results to forecast customers' behavior in advance (Custer, 2021). Therefore, helping customer support services offer the most likely service option or information. For instance, Netflix, an American Over-The-Top (OTT) streaming service, uses AI to understand the customer behavior pattern regarding the type of content watched and the country of residence. Using these two data points in tandem, the customer support team of Netflix personalizes the feed for each customer, whereby efforts are being made to personalize the movie's thumbnail or show for each customer (Chandrashekar et al., 2017). Thus, for the film *Goodwill Hunting*, Netflix will use a romantic thumbnail of the lead actors for the customer who has a history of watching romantic movies. Alternatively, the Netflix AI will show a thumbnail of the comedian from the film to the customer watching comedy movies or shows in the past (Owen, 2022). Thus, Netflix uses AI to analyze customer behavior patterns and provide personalized suggestions to improve customer engagement. This analysis and forecast of customer behavior patterns enable AI to address customers' issues instantly. Hence, it leads to the next element of functional AI, issue addressal.

Issue Addressal

In addition to analyzing customer behavior patterns, AI-powered tools can solve customer issues effectively and efficiently. AI-enabled means such as chatbots, robots, IVRs, and so on can work 24/7/365 without getting tired (Duggal, 2022). This unique capability of AI and the ability to tag every customer-related information based on rules make them less prone to errors and bias (Roldós, 2020). This consistency and accuracy of AI-enabled algorithms help customer support services rely on insights from AI in making decisions (McColl-Kennedy et al., 2019). Furthermore, the AI interfaces can help perform lower-level tasks such as answering phone calls, providing answers to FAQs, and even approving refunds without human intervention. Therefore, assisting the customer support teams in resolving customer issues independently and at any hour of the day (Perrotta, 2019). For instance, the AI-enabled chatbot TOBi from Vodafone is one of the most prominent applications of AI in the telecom industry. TOBi operates in 15 countries worldwide (Donegan, 2021) and helps customer support services handle 100% of the messaging conversations received through text or voice commands. TOBi also helps resolve 70% of customer issues independently and instantly (IBM, 2021). Hence, assisting Vodafone in enhancing the customer experience by providing solutions to customer queries around the clock. Thus, AI facilitates customer support services in addressing customer-related issues 24/7/365. Furthermore, AI helps solve customers' problems instantly, reducing the response time and enhancing the customer experience.

Quick Response Time

Along with solving customer issues, AI uses its intelligence to reduce the response time by automatically sorting customer issues based on several factors such as subject matter, language, sentiment, and so on and preparing the trajectory to route these issues to the right person at the right time (Edlich et al., 2019). Identifying the problems, classifying them into different stacks, and routing them to the relevant person helps improve response time and deliver an effective and efficient customer experience across multiple channels. For instance, the customers of one of the biggest banks, Yapi Kredi in Turkey, wanted a better dial-up facility for customer support services. The dial-up process was lengthy as the customer had to listen to the full menu before navigating across the menu. This process demanded too much customer effort, thus leading to a negative customer experience. However, the deployment of the interactive voice response (IVR) system using AI helped the bank solve this problem by redirecting the customers to the relevant menu by using voice commands. As a result, the IVR system started routing three million voice calls monthly, reducing the call length and customer efforts by 65%. Hence, drastically reducing the time required to respond to customers. As a result, Yapi Kredi saw an increase in customers' overall satisfaction by 10% (Sestek, 2021). Hence, the reduction in response time leads to the enhancement of customer experience. Furthermore, the ability of AI to anticipate customers' sentiments helps enhance customer experience manifold.

Anticipating Sentiments

The AI-enabled algorithms can anticipate customer sentiments from the customers' words, sentences, and voice commands (Mesevage, 2020). The basic AI follows a predetermined metric to understand and classify customer sentiments as positive, negative, or neutral. Over time, AI learns and begins to identify more complex feelings, such as sarcasm (Alkhaldi, 2021). Anticipating customer sentiments helps extract actionable information from unstructured customer data. Therefore, it enables customer support services to anticipate customer sentiments at the moment and react accordingly. For instance, a research team in Malaysia used AI to understand customers' emotions toward services provided by public hospitals. The research used Facebook reviews to anticipate customers' sentiments across five service dimensions: staff empathy, assurance, responsiveness, reliability, and tangibility. The results showed an optimistic view for each of the five dimensions. However, this research helped the hospitals in Malaysia to further improve on dimensions with low scores, such as empathy and reliability, to further enhance customer experience (Mande, 2022; Rahim et al., 2021). Hence, AI helps customer service providers analyze the efforts' results and identify areas for improvement. Therefore, leading to the creation of an AI-enabled ecosystem that enhances customer experience. However, AI should be used with ethics to gain customer trust and build strong relationships. The following section thus discusses various elements related to ethical AI.

Ethical AI

As discussed in the previous section, the adoption and use of AI in customer support services are helping customers receive solutions to problems. However, this advancement in the use of AI by customer support services poses a challenge for businesses regarding the ethical use of AI. The ethical use of AI refers to the challenge of building trust between the company and the customer related to the information used by AI (Herbert, 2022). For example, an AI-based chatbot should let customers know they are interacting with a machine, not a human (Etlinger, 2017). Similarly, businesses should focus on letting customers know about the limitations of bots and information regarding the use of sensitive information (Clark, 2021a). Therefore, using AI in customer support services becomes ethical with four key pillars: educating customers, inclusivity, transparency, and empowering customers. The following sub-sections will discuss the four key pillars in detail.

Educate the Customer

The first pillar of ethical AI involves educating the customers regarding using AI in customer support services. The customer support teams can educate customers by first making them aware of the use of AI-enabled algorithms in chatbots and IVR and, second, making them aware of the capabilities and limitations of AI (Marr, 2021). This process of educating customers helps set up their expectations of customers. For instance, Trip Advisor, an American online travel and tourism brand, educates customers regarding the virtual travel agent (AI-enabled chatbot), collecting basic information about the trip. This brand initiative helps customers know whom they are interacting with and what can be expected. Hence, educating customers in advance leads to setting the right expectations. Therefore, enabling the brand to enhance the overall customer experience (Menze, 2018). In addition to educating the customers, customer support services need to be transparent regarding the use of customer data. The following sub-section discusses the importance of maintaining transparency in detail.

Transparent

The second pillar of ethical AI involves transparency. The customer support services are expected to be open and honest regarding the use of customer data. The relevant and correct information needs to be shared with the customers based on data collected, usage, and saved data. The openness in sharing such critical information helps the firms build customer trust. For instance, the Royal Bank of Scotland, headquartered in Edinburgh, UK, prefers sharing every minute detail related to capturing, using, and disseminating customer data by AI. This transparency in customer data helps the bank build strong customer relationships. As a result, customers feel connected with the bank and become comfortable sharing more data to get personalized experiences (Marr, 2021). Hence, this leads us to the third pillar of ethical AI, inclusivity. The influence of inclusivity on customer experience is discussed as follows.

Inclusive

The third pillar of ethical AI revolves around creating an inclusive system through AI. White men dominate the teams that develop AI algorithms (Marr, 2021). This domination reflects biases toward one segment of society in the results. For instance, Goldman Sach was sacked for discriminating between men and women in its credit card practices. The issue came to light when an entrepreneur and software developer from Denmark tweeted about having a 20 times more credit limit on an Apple card than his wife, even when she had a higher credit score (Telford, 2019). Hence, efforts should be made to involve people from different gender, caste, creed, color, and geographical locations to build an inclusive customer support system through AI. Furthermore, random checks on AI-enabled algorithms should also be conducted to rule out biases and promote inclusivity. Therefore, inclusive AI paves the way to empower customers by providing them with the equality and voice to take charge of their journey.

Empower Customer

The fourth pillar of ethical AI involves taking measures to empower customers to take charge of their customer journey. The customers feel empowered when they can choose their path for each interaction with customer support services. Therefore, customer support services need to build a system that allows customers flexibility regarding accessing information. For example, AI-driven WhatsApp Business chatbot will enable customers to connect with customer support services through instant messaging. However, this application empowers customers by allowing them to continue the chat or opt out with a click. This feature of WhatsApp Business empowers customers by giving them the freedom to make choices (Jain, 2020). Therefore, customers can fully benefit from customer support services by designing their journey. Similarly, the involvement of humans at different levels of the customer journey helps in cocreating the customer experience.

Human Interaction

The increased use of functional and ethical AI in customer support services requires human interaction at all stages to cocreate customer experience (Huang & Rust, 2018). While AI is growing at different stages of customer support services, it should not be mistaken as a replacement for human interaction (Hermann, 2022). AI is an enabler that compliments the customer support service providers by taking care of functional issues.

For example, AI helps customer support service providers by answering regular customer queries and redirecting more critical and complex issues to the required team (Adamopoulou & Moussiades, 2020). This support from AI helps customer service providers enhance customers' overall experience. Thus, AI serves as an enabler to customer support service providers to improve the overall customer experience. For instance, the Royal Caribbean cruise, headquartered in Miami, Florida, has built and deployed an onboard chatbot as a virtual concierge and employed natural language processing (NLP) and machine learning. This chatbot is designed with a human-in-the-loop model whereby the chatbot sends a pager to the required customer service provider to address issues beyond the chatbot's capabilities (Marr, 2019). Furthermore, the data collected by the chatbot gets integrated with the brands' customer relationship management (CRM) to ensure an easy and convenient flow of information to derive solutions for customers' problems.

Integrated CRM

Customer support service providers use AI and customer relationship management (CRM) to pave the way for a bright future for customer support services (Ledro et al., 2022). AI, CRM, and human interaction help automate functional tasks, providing relevant data and information to service providers to swiftly address customer problems (MIT Technology Review Insights, 2020). For example, Sunbasket, a subscription-based meal delivery service in San Francisco, uses Salesforce's AI-powered CRM system Einstein to automate functional tasks such as determining the nature of a customer call and directing it to the required department. This automated system facilitates Sunbasket's service providers in quickly addressing customer queries, therefore helping them build trust and enhance customer experience (Salesforce., 2020). The trust of the customers is further elevated by ensuring the maintenance of customer data privacy.

Privacy

AI in customer support services involves collecting a considerable amount of customer data. The collection and processing of this data pose a risk of a breach of customers' privacy which is also mentioned by the latest report from TechRepublic; 86% of the customers feel concerned about data privacy. Additionally, 40% reported needing more trust in customer support services regarding the ethical use of personal data (Clark, 2021b). These statistics reveal the customers' concerns regarding data privacy while using customer support services. Hence, customer support teams are expected to take action to protect customers from breach of confidentiality. The customer support services use encryption, Blockchain, and federated learning while using AI to safeguard customers' privacy. These components related to protecting customer privacy are discussed in detail in the following subsections.

Encryption

The first step toward safeguarding customer data involves encrypting personal details in random texts before saving them in CRM systems. The encryption of customer data ensures privacy as specific customer information is encrypted and scrambled to random texts (Rafter, 2022). Hence, this encryption process helps customer support services safeguard customer data and prevent breaches of customer privacy. For instance, the Marine Corps, the Defense Travel System (DTS) of the US 2018, sent out an unencrypted mail containing the personal information of approximately 21,500 civilians to the wrong mailing list. As a result, the private and sensitive information related to officers' bank account numbers, social security numbers, and emergency contact information was shared with unknown people. This incident resulted in the US government's backlash against the Marine Corps. Post this incident, the Marine Corps has used AI to encrypt and protect sensitive information related to civilians. DTS has also limited the maximum number of email ids that can be included while sending group mail (Bisson, 2020). Hence, taking precautionary measures to safeguard customers' privacy. Additionally, customer service providers use Blockchain to avoid breaches of customer privacy while transferring data from the point of collection to the point of dissemination.

Blockchain

In addition to encryption, customer support services use AI-enabled Blockchain to securely transfer encrypted data from one department to another. The AI-enabled Blockchain can be defined as an immutable ledger facilitating the instant, secured, and transparent exchange of encrypted data among different customer support service departments to provide a complete solution to the problem (Cuomo, 2021). For instance, IPwe, the world's first AI and Blockchain-powered patent platform, created an automated marketplace that facilitates the connection between all the parties involved in creating and using the patent. This platform is known as Global Patent Registry (GPR), and all the information is shared using AI and Blockchain. Therefore, helping GPR maintain the privacy of customer data while sharing the information with the relevant client (King, 2021).

Protect Sensitive Data

Using AI-enabled encryption and Blockchain by customer support services builds customer trust with privacy and sharing of sensitive data. In a recent study by PwC, 63% of the customers reported sharing private and sensitive data with the brands they trust. However, 43% refused to share personal data with brands they did not trust (Puthiyamadam & Reyes, 2021). Hence, the customer support services of brands need to protect sensitive data using AI to build customer trust. Eighty-eight percent of Americans reported trust as the #1 factor for sharing personal data (Marr, 2021). However, a failure to protect sensitive data can lead to negative results. For instance, Facebook faced a significant backlash when the company shared the personal data of over 50 million customers with Cambridge Analytica (Wong, 2019). As a result of this breach of transparency, Facebook faced several consequences, such as drop-in customers and a 20% reduction in customer engagement (Hern, 2019). The drop in customers resulted in distrust between the brand and the customer. Hence, customer support services invest in federated learning, an additional method to safeguard customer privacy and help customer support services enhance the customer experience.

Federated Learning

In addition to encryption and Blockchain, federated learning is an additional way of enhancing customer privacy in customer support services. Federated learning is the intelligent application of AI that builds new models on existing data from different sources without transferring original information (Clark, 2021b). In other words, federated learning works on building models on different customer devices independently by utilizing the on-device data. These models are then shared across devices to derive deeper insights without sharing actual data (Gupta, 2019). Hence, federated learning helps customer support service providers maintain customers' privacy while drawing deeper insights. For instance, Google's Android Keyboard uses federated learning to independently improvise predictive texts for the customer on each device and collectively collaborate the results to enhance the overall experience (Taylor, 2021). Hence, paving the way to personalize each customer's experience while maintaining their privacy.

Personalized Customer Services

AI-enabled algorithms facilitate customer support services in collecting and analyzing customers' historical data within seconds. The AI further helps detect customer behavior patterns and sentiments to personalize the experience. This personalization of experience helps influence customer decisions related to purchasing (73%). Furthermore, a recent report from PwC found that 43% of all customers agreed to pay a higher price for convenience and 42% for a friendly and welcoming experience

(Puthiyamadam & Reyes, 2021). Hence, personalization in customer experience can serve as a game changer for businesses acquiring and retaining customers. For instance, Marriott International—a luxury hotel chain with hotels worldwide—uses multiple AI-enabled chatbots (Agarwal, 2018). These multilingual chatbots are available on various social media platforms such as Facebook, WeChat, Slack, Google Assistant, and so on (Mest, 2017).

Furthermore, these NLP-aided chatbots help detect customer behavior and preferences to provide a hyper-personalized experience. Therefore, taking care of every need of the customer with minute details. As a result, the hotel chain has earned the highest number of satisfied and repeat customers in the luxury hotel industry (Bethesda, 2017). Hence, using AI with ethics and privacy helps customer support services enhance the customers' personalized experience.

Implications to Academia and Industry

Using AI with ethics and privacy in customer support services has several implications for academia and industry. We have listed two of these implications in the following subsections:

Academic Implications

This chapter advances the field of customer support services by providing the following three academic implications:

The chapter proposes a detailed framework to map the components involved in delivering personalized customer services while maintaining ethical and privacy boundaries for customers using AI-enabled tools.

The development of the proposed framework enabled a critical review of the existing literature around the key components of the framework in the multi-disciplinary space of marketing and information technology.

The relationship between the proposed framework components—using AI with ethics and privacy in customer support services (see

Fig. 10.1)—can be quantitatively tested, and suitable moderators and mediators can be included to enhance the proposed model.

Practical Implications

The growing importance of customer support services in businesses makes the advancement of this department crucial. Hence, we propose the following recommendations to customer support services providers, AI developers, and marketing managers of companies to enhance their customer support services.

Using AI and other advanced technology in customer support services should not be treated as a replacement for humans; instead, it should be treated as a partner to humans. Hence, the customer support service providers and AI developers should work as a team to improve customer experiences. Therefore, efforts must be laid to divide responsibilities between AI-enabled tools and customer service providers. The devices should be responsible for handling functional tasks, and customer support service providers should intervene when AI-enabled tools fail to address a specific or complex query.

AI in customer support services helps enhance customers' personalized experience. However, a continuous check by AI developers and marketing managers to assess the impact of AI on customers should be carried out to minimize friction, disappointments, and frustration.

The AI algorithms are self-learning tools. These tools learn to improve their performance and provide a seamless experience to customers. However, AI-enabled algorithms can work when customers share their data with customer support services. Hence, customer support services must invest in educating their customers regarding the protection of customer privacy to build trust in customers and encourage them to share personal data.

Conclusion

This chapter contributes to using AI with ethics and privacy in customer support services. Customer support services use AI to learn about customers' behavior patterns, address customer issues, reduce response time, and anticipate sentiments to provide personalized services. Additionally, using AI in customer support services further helps brands educate and empower customers by being transparent and inclusive. The empowered customers make their own choices and assist customer support services in personalizing the experience. However, customer support services require customers to share personal data to personalize. The sharing of personal data raises doubts regarding the breach of privacy. Hence, customer support services use encryption, Blockchain, and federated learning to maintain customer data privacy. Privacy concerns are further addressed by continuous human intervention. Therefore, the new-age customer support services work on cocreation between technology and human. This cocreation of customers and humans leads to the enhancement of customer experience. Thus, paving the way for experiments with technology to enhance customer experiences. This advancement also opens up opportunities to integrate marketing and IT departments in firms to help manage work collectively toward providing a personalized and seamless customer experience.

References

Adamopoulou, E., & Moussiades, L. (2020). Chatbots: History, technology, and applications. *Machine Learning with Applications, 2*, 100006. https://doi.org/10.1016/j.mlwa.2020.100006

Agarwal, N. (2018, October 5). Top 5 ways Chatbots are revolutionizing the hotel industry. *Medium*. https://chatbotslife.com/top-5-ways-chatbots-are-revolutionizing-the-hotel-industry-a3a07e7ea895

Alkhaldi, N. (2021, November 25). AI-driven sentiment analysis in customer service: Use cases, benefits, Expert Tips. *ITRex*. https://itrexgroup.com/blog/ai-sentiment-analysis-in-customer-service-use-cases-benefits-expert-tips/

Ameen, N., Tarhini, A., Reppel, A., & Anand, A. (2021). Customer experiences in the age of artificial intelligence. *Computers in Human Behavior, 114*, 106548. https://doi.org/10.1016/j.chb.2020.106548

Bélanger, F., & Crossler, R. E. (2011). Privacy in the digital age: A review of information privacy research in information systems. *MIS Quarterly, 35*(4), 1017. https://doi.org/10.2307/41409971

Bethesda. (2017, September 28). Marriott International's AI-powered Chatbots on Facebook messenger and slack and Aloft's ChatBotlr, simplify travel for guests throughout their journey. *Marriott International Newscenter (US).* https://news.marriott.com/news/2017/09/28/marriott-internationals-ai-powered-chatbots-on-facebook-messenger-and-slack-and-alofts-chatbotlr-simplify-travel-for-guests-throughout-their-journey

Bisson, D. (2020, October 15). 7 Data breach case studies caused by human error. *Venafi.* https://www.venafi.com/blog/7-data-breaches-caused-human-error-did-encryption-play-role

Bock, D. E., Wolter, J. S., & Ferrell, O. C. (2020). Artificial intelligence: Disrupting what we know about services. *Journal of Services Marketing, 34*(3), 317–334. https://doi.org/10.1108/JSM-01-2019-0047

Chandrashekar, A., Amat, F., Basilico, J., & Jebara, T. (2017, December 7). Artwork personalization at Netflix | by Netflix technology blog | Netflix TechBlog. *Netflix.* https://netflixtechblog.com/artwork-personalization-c589f074ad76

Chatterjee, S., Chaudhuri, R., & Vrontis, D. (2022). AI and digitalization in relationship management: Impact of adopting AI-embedded CRM system. *Journal of Business Research, 150*, 437–450. https://doi.org/10.1016/j.jbusres.2022.06.033

Clark, S. (2021a). *What is ethical AI and why is it vitally important?* https://www.cmswire.com/digital-experience/what-is-ethical-ai-and-why-is-it-vitally-important/.

Clark, S. (2021b, November 4). How AI is helping to better protect customer data and privacy. *CMSWire.Com.* https://www.cmswire.com/customer-experience/how-ai-is-being-used-to-protect-customer-privacy/.

Costello, K. (2019, January 21). Gartner survey shows 37 percent of organizations have implemented AI in some form. *Gartner.* https://www.gartner.com/en/newsroom/press-releases/2019-01-21-gartner-survey-shows-37-percent-of-organizations-have

Cuomo, J. (2021). Blockchain and artificial intelligence (AI) | *IBM.* https://www.ibm.com/in-en/topics/blockchain-ai.

Custer, C. (2021, May 27). A complete guide to customer behavior analysis in 2022. *Woopra*. https://www.woopra.com/learn/customer-behavior-analysis

Davenport, T., Guha, A., Grewal, D., & Bressgott, T. (2020). How artificial intelligence will change the future of marketing. *Journal of the Academy of Marketing Science, 48*(1), 24–42. https://doi.org/10.1007/s11747-019-00696-0

Davenport, T., & Ronanki, R. (2018). *Artificial intelligence for the real world*. https://hbr.org/2018/01/artificial-intelligence-for-the-real-world.

DeMaio, P. (2020, June 22). *How COVID-19 has changed the way customer service leaders think about contact centers*. https://aws.amazon.com/blogs/contact-center/how-covid-19-has-changed-the-way-customer-service-leaders-think-about-contact-centers/

Diener, E. (1984). Subjective well-being. *Psychological Bulletin, 95*(3), 542–575.

Dignum, V. (2018). Ethics in artificial intelligence: Introduction to the special issue. *Ethics and Information Technology, 20*(1), 1–3. https://doi.org/10.1007/s10676-018-9450-z

Doll, T. (2019, November 24). A machine learning approach to automated customer satisfaction surveys. *Medium*. https://towardsdatascience.com/a-machine-learning-approach-to-automated-customer-satisfaction-surveys-946d2604e309.

Donegan, M. (2021, October 13). DTWS: Vodafone shares how TOBi is growing up. *TM Forum Inform*. https://inform.tmforum.org/ai-data-and-insights/2021/10/dtws-vodafone-shares-how-tobi-is-growing-up/.

Duggal, N. (2022, August 1). *Advantages and disadvantages of artificial intelligence. Simplilearn.com*. https://www.simplilearn.com/advantages-and-disadvantages-of-artificial-intelligence-article.

Dwivedi, Y. K., Hughes, L., Ismagilova, E., Aarts, G., Coombs, C., Crick, T., Duan, Y., Dwivedi, R., Edwards, J., Eirug, A., Galanos, V., Ilavarasan, P. V., Janssen, M., Jones, P., Kar, A. K., Kizgin, H., Kronemann, B., Lal, B., Lucini, B., et al. (2021). Artificial intelligence (AI): Multi-disciplinary perspectives on emerging challenges, opportunities, and agenda for research, practice and policy. *International Journal of Information Management, 57*, 101994. https://doi.org/10.1016/j.ijinfomgt.2019.08.002

Edlich, A., Phalin, G., Jogani, R., & Kaniyar, S. (2019). Driving impact at scale from automation and AI. *McKinsey & Company*. https://www.mckinsey.com/~/media/McKinsey/Business%20Functions/McKinsey%20Digital/Our%20Insights/Driving%20impact%20at%20scale%20from%20automation%20and%20AI/Driving-impact-at-scale-from-automation-and-AI.ashx.

Ertel, W. (2018). *Introduction to artificial intelligence*. Springer.

Etlinger, S. (2017, November 17). *Ethics will shape the customer experience of AI.* https://www.linkedin.com/pulse/ethics-shape-customer-experience-ai-susan-etlinger/.

Forbes. (2021, July 22). Council post: 15 ways to leverage AI in customer service. *Forbes.* https://www.forbes.com/sites/forbesbusinesscouncil/2021/07/22/15-ways-to-leverage-ai-in-customer-service/

Gladly. (2018, June 27). Customer service expectations survey and statistics. *Gladly.* https://www.gladly.com/blog/2018-customer-service-expectations-survey/.

Gupta, A. (2019, December 11). How federated learning is going to revolutionise AI. *Medium.* https://towardsdatascience.com/how-federated-learning-is-going-to-revolutionize-ai-6e0ab580420f.

Hagendorff, T. (2020). The ethics of AI ethics: An evaluation of guidelines. *Minds and Machines, 30*(1), 99–120. https://doi.org/10.1007/s11023-020-09517-8

Haptik. (2021, January 27). *How Chatbots are poised to transform customer service.* https://www.haptik.ai/blog/customer-service-chatbots

Herbert, D. (2022, March 10). Responsibility and AI ethics in the contact centre. *Odigo EN.* https://www.odigo.com/en-gb/blog-and-resources/blog/the-10-most-revealing-customer-experience-stats-from-2020/.

Hermann, E. (2022). Leveraging artificial intelligence in marketing for social good—An ethical perspective. *Journal of Business Ethics, 179*(1), 43–61. https://doi.org/10.1007/s10551-021-04843-y

Hern, A. (2019, June 20). Facebook usage falling after privacy scandals, data suggests. *The Guardian.* https://www.theguardian.com/technology/2019/jun/20/facebook-usage-collapsed-since-scandal-data-shows

Heskett, J. L. (1962). Ferment in marketing's oldest area. *Journal of Marketing, 26*(4), 40. https://doi.org/10.2307/1248337

Huang, M.-H., & Rust, R. T. (2018). Artificial Intelligence in Service. *Journal of Service Research, 21*(2), 155–172. https://doi.org/10.1177/1094670517752459

IBM. (2021, August 17). *Vodafone GmbH.* https://www.ibm.com/case-studies/vodafone

Jain, S. (2020, September 11). WhatsApp business Chatbot with AI: What are the benefits and how do I create one? *ORAI Robotics.* https://www.orai-robotics.com/post/whatsapp-business-chatbot-with-ai-what-are-the-benefits-and-how-do-i-create-one

Kaplan, A., & Haenlein, M. (2019). Siri, Siri, in my hand: Who's the fairest in the land? On the interpretations, illustrations, and implications of artificial intelligence. *Business Horizons, 62*(1), 15–25. https://doi.org/10.1016/j.bushor.2018.08.004

Kautish, P., & Khare, A. (2022). Investigating the moderating role of AI-enabled services on flow and awe experience. *International Journal of Information Management, 66*, 102519. https://doi.org/10.1016/j.ijinfomgt.2022.102519

Kim, W., Ryoo, Y., Lee, S., & Lee, J. A. (2022). Chatbot advertising as a double-edged sword: The roles of regulatory focus and privacy concerns. *Journal of Advertising, 1*, 1–19. https://doi.org/10.1080/00913367.2022.2043795

King, B. (2021, May 5). AI and Blockchain help discover and transact IP. *IBM*. https://www.ibm.com/case-studies/ipwe.

Kleinings, H. (2022, July 26). *AI for customer support and why you need it*. https://levity.ai/blog/ai-for-customer-support.

Krayewski, K. (2022, April 13). *Will AI replace humans in customer service?* https://www.ultimate.ai/blog/customer-experience/will-ai-replace-humans-in-customer-service

Ledro, C., Nosella, A., & Vinelli, A. (2022). Artificial intelligence in customer relationship management: Literature review and future research directions. *Journal of Business & Industrial Marketing, 37*(13), 48–63. https://doi.org/10.1108/JBIM-07-2021-0332

Lee, D., Sirgy, J., Larsen, V., & Wright, N. (2002). Developing a subjective measure of consumer well-being. *Journal of Macromarketing, 22*(2), 158–169. https://doi.org/10.1177/0276146702238219

Lusch, R. F., & Vargo, S. L. (2006). Service-dominant logic: Reactions, reflections and refinements. *Marketing Theory, 6*(3), 281–288. https://doi.org/10.1177/1470593106066781

Lusch, R. F., & Vargo, S. L. (2011). Service-dominant logic: A necessary step. *European Journal of Marketing, 45*(7/8), 1298–1309. https://doi.org/10.1108/03090561111137723

Mande, P. (2022, January 31). *A guide to empathy in customer service + empathy statements to use*. https://freshdesk.com/customer-service-skills/guide-empathy-customer-service-blog/

Marr, B. (2019, May 10). AI on cruise ships: The fascinating ways Royal Caribbean uses facial recognition and machine vision. *Forbes*. https://www.forbes.com/sites/bernardmarr/2019/05/10/the-fascinating-ways-royal-caribbean-uses-facial-recognition-and-machine-vision/

Marr, B. (2021, September 10). How do we use artificial intelligence ethically? *Forbes*. https://www.forbes.com/sites/bernardmarr/2021/09/10/how-do-we-use-artificial-intelligence-ethically/

McColl-Kennedy, J. R., Zaki, M., Lemon, K. N., Urmetzer, F., & Neely, A. (2019). Gaining customer experience insights that matter. *Journal of Service Research, 22*(1), 8–26. https://doi.org/10.1177/1094670518812182

Menze, J. (2018, April 17). How TripAdvisor uses AI to "humanize" its review community | *PhocusWire*. https://www.phocuswire.com/TripAdvisor-AI-reviews

Mesevage, T. G. (2020, October 7). AI-enabled sentiment analysis in business. *MonkeyLearn Blog*. https://monkeylearn.com/blog/ai-sentiment-analysis/.

Mest, E. (2017, October 26). Marriott launches Facebook chatbot to assist in career development | *Hotel Management*. https://www.hotelmanagement.net/human-resources/marriott-launches-facebook-chatbot-to-assist-career-development

MIT Technology Review Insights. (2020, September 17). From support function to growth engine: The future of AI and customer service. *MIT Technology Review*. https://www.technologyreview.com/2020/09/17/1008148/from-support-function-to-growth-engine-the-future-of-ai-and-customer-service/.

Mogaji, E., Soetan, T. O., & Kieu, T. A. (2021). The implications of artificial intelligence on the digital marketing of financial services to vulnerable customers. *Australasian Marketing Journal, 29*(3), 235–242. https://doi.org/10.1016/j.ausmj.2020.05.003

Mökander, J., & Floridi, L. (2021). Ethics-based auditing to develop trustworthy AI. *Minds and Machines, 31*(2), 323–327. https://doi.org/10.1007/s11023-021-09557-8

Morley, J., Kinsey, L., Elhalal, A., Garcia, F., Ziosi, M., & Floridi, L. (2021). Operationalising AI ethics: Barriers, enablers and next steps. *AI & Society, 38*, 411. https://doi.org/10.1007/s00146-021-01308-8

Onorato, A. (2022, May 3). *Consumers open to AI in marketing, but data privacy matters*. https://cdp.com/articles/report-consumers-open-to-ai-in-marketing-but-privacy-concerns-remain/.

Owen, R. (2022, January 10). Artificial intelligence at Netflix—Two current use-cases. *Emerj Artificial Intelligence Research* https://emerj.com/ai-sector-overviews/artificial-intelligence-at-netflix/

Perrotta, S. T. (2019, April). How AI is reducing handle time and improving the customer experience. *Contact Center Pipeline*. https://www.contactcenterpipeline.com/Article/how-ai-is-reducing-handle-time-and-improving-the-customer-experience.

Pine, J., & Gilmore, J. H. (1998). Welcome to the experience economy. *Harvard Business Review, 76*, 97. https://hbr.org/1998/07/welcome-to-the-experience-economy

Puntoni, S., Reczek, R. W., Giesler, M., & Botti, S. (2021). Consumers and artificial intelligence: An experiential perspective. *Journal of Marketing, 85*(1), 131–151. https://doi.org/10.1177/0022242920953847

Puthiyamadam, T., & Reyes, J. (2021). Experience is everything: Here's how to get it right. *PwC*. https://www.pwc.com/us/en/services/consulting/library/consumer-intelligence-series/future-of-customer-experience.html

Rafter, D. (2022, March 15). *What is encryption and how does it protect your data?* https://us.norton.com/internetsecurity-privacy-what-is-encryption.html

Rahim, A. I. A., Ibrahim, M. I., Musa, K. I., Chua, S.-L., & Yaacob, N. M. (2021). Assessing patient-perceived hospital service quality and sentiment in Malaysian public hospitals using machine learning and Facebook reviews. *International Journal of Environmental Research and Public Health, 18*(18), 9912. https://doi.org/10.3390/ijerph18189912

Roldós, I. (2020, March 27). How is AI revolutionizing customer service? *MonkeyLearn Blog*. https://monkeylearn.com/blog/customer-care-ai/

Rowan, M. (2019, September 17). How Singapore Airlines uses AI to personalize air travel. *KPItarget*. http://www.kpitarget.com/how-singapore-airlines-uses-ai-to-personalize-air-travel/.

Salesforce. (2020, June 24). Growth, innovation & giving back – Sun Basket's response to COVID-19. *Salesforce News*. https://www.salesforce.com/news/stories/growth-innovation-s-response-to-covid-19/

Schneider, C. (2017, October 16). 10 reasons why AI-powered, automated customer service is the future. *Watson Blog*. https://www.ibm.com/blogs/watson/2017/10/10-reasons-ai-powered-automated-customer-service-future/

Sestek. (2021). Conversational IVRYapı Kredi. *SESTEK*. https://www.sestek.com/case-studies/yapi-kredi-sestek-conversational-ivr-case-study/

Sheth, J. (2020). Impact of Covid-19 on consumer behaviour: Will the old habits return or die? *Journal of Business Research, 117*, 280–283. https://doi.org/10.1016/j.jbusres.2020.05.059

Sheth, J., Jain, V., & Ambika, A. (2020). Repositioning the customer support services: The next frontier of competitive advantage. *European Journal of Marketing, 54*(7), 1787–1804. https://doi.org/10.1108/EJM-02-2020-0086

Sheth, J. N. (2011). Impact of emerging markets on marketing: Rethinking existing perspectives and practices. *Journal of Marketing, 75*(4), 166–182. https://doi.org/10.1509/jmkg.75.4.166

Siau, K., & Wang, W. (2020). Artificial intelligence (AI) ethics: Ethics of AI and ethical AI. *Journal of Database Management, 31*(2), 74–87. https://doi.org/10.4018/JDM.2020040105

Simon, L. S. (1965). Measuring the market impact of technical services. *Journal of Marketing Research, 2*(1), 32. https://doi.org/10.2307/3149334

Spencer-Matthews, S., & Lawley, M. (2006). Improving customer service: Issues in customer contact management. *European Journal of Marketing, 40*(1/2), 218–232. https://doi.org/10.1108/03090560610637392

Taylor, K. (2021). 5 applications of federated learning. *HitechNectar*. https://www.hitechnectar.com/blogs/applications-of-federated-learning/.

Telford, T. (2019, November 11). Apple Card algorithm sparks gender bias allegations against Goldman Sachs. *Washington Post*. https://www.washingtonpost.com/business/2019/11/11/apple-card-algorithm-sparks-gender-bias-allegations-against-goldman-sachs/.

Trivedi, M. (2022). 10 ways artificial intelligence can improve customer service. *TA Digital*. https://www.tadigital.com/insights/perspectives/10-ways-artificial-intelligence-can-improve-customer-service.

Velazquez, R. (2022, July 28). Artificial intelligence. What is artificial intelligence? How does AI work? https://builtin.com/artificial-intelligence

Violino, B. (2021, May 24). How AI can enhance customer experience. *InfoWorld*. https://www.infoworld.com/article/3618173/how-ai-can-enhance-customer-experience.html

Wong, J. C. (2019, March 18). The Cambridge Analytica scandal changed the world – But it didn't change Facebook. *The Guardian*. https://www.theguardian.com/technology/2019/mar/17/the-cambridge-analytica-scandal-changed-the-world-but-it-didnt-change-facebook

Writer, M. S. (2019, August 27). 20 million fliers to experience Singapore airlines' AI-powered experiences. *MarTech Series*. https://martechseries.com/predictive-ai/ai-platforms-machine-learning/20-million-fliers-experience-singapore-airlines-ai-powered-experiences/

Zhu, Y., Zhang, J., Wu, J., & Liu, Y. (2022). AI is better when I'm sure: The influence of certainty of needs on consumers' acceptance of AI chatbots. *Journal of Business Research, 150*, 642–652. https://doi.org/10.1016/j.jbusres.2022.06.044

11

Linking Customer E-Service Quality with Artificial Intelligence-Based Business Environment

Sakshi Kathuria and Sudhir Rana

Introduction

Many technological advancements make customer experiences comfortable in the Fourth Industrial Revolution. This technological advancement refers to artificial intelligence (AI), blockchain technology (BCT), industrial internet of things (IIoT), and big data analytics (BDA) that changed the entire customer experience. (Bolton et al., 2018b; Mogaji et al., 2020). These technologies change not only the customer experiences but provide a competitive edge for business organizations to increase the business outcome in the digital disruption era (Rouhani et al., 2016; Wamba-Taguimdje et al., 2020) by facilitating the customers' product

S. Kathuria (✉)
Fortune Institute of International Business (FIIB), New Delhi, India
e-mail: sakshi.kathuria@fiib.edu.in

S. Rana
College of Healthcare Management and Economics,
Gulf Medical University, Ajman, UAE

© The Author(s), under exclusive license to Springer Nature Switzerland AG 2023

J. N. Sheth et al. (eds.), *Artificial Intelligence in Customer Service*,
https://doi.org/10.1007/978-3-031-33898-4_11

and service offerings (Balaji & Roy, 2017; Mogaji & Nguyen, 2022; Wang & Fan, 2020). AI is a widely used emerging technology that helps organizations track the customer's real-time responses to meet the requirement in the future time and share a predictive environment that helps retailers meet the consumer's expectations. AI also offers an ecosystem that records consumer behavior as an essential tool for measuring the consumers' service quality in terms of customer attraction and retention. The consumer's buying behavior and expected move with overall experience are recorded with the multiple AI-enabled tools. Also, AI tools are used to deduce customer expectations and move forward to improve service quality (Balaji & Roy, 2017).

AI is based on a series of technologies that enable a cognitive system to think, understand, react, and consider feedback to improve customer interactions (Bowen & Morosan, 2018). Artificial intelligence finds its wide application in various products and services business environments and provides sustainable and digital edges to customer experiences. With the advancement of digital technology, entire global business organizations are connected over an interconnected network. It requires different levels of intelligence to control the product and service flow, reduce task complexity and usage intention, and support a problem-solving ability for business growth (Huang & Rust, 2021; Xu et al., 2020a). The smart technologies implementation leads to a human and resources investment in AI for handling mega data and predicting the market requirement and expectations. AI has wider applications in e-commerce, medicine, law, education, and manufacturing. However, the current consumer-driven market scenario requires the connection of AI-driven consumer services that help to improve consumer well-being in terms of trust in the technology, commitment toward the product, and loyalty to the brand based on the program-based algorithm (Davenport et al., 2020). Unlike human intelligence, AI is a machine-controlled intelligence that processes heterogeneous consumer data simultaneously to tackle data complexity. Russel and Norvig (2012) defined AI as a machine that can simulate the human brain's various cognitive and affective functions much faster. However, with the help of various data-controlled machines, workers never have to experience a similar job as boring. An artificially intelligent system can perform a repetitive task (Castillo et al., 2021). The AI network and critical consumer value chain can be established through AI and machine learning tools to support consumer well-being (Li et al., 2021). Customer experiences can be improved through AI-driven chatbots with natural language processing (NLP) (Abdulquadri et al., 2021; Balakrishnan et al., 2021; Russel & Norvig, 2012).

The identification and analysis of customer habits, purchase intention, product trust, and feedback on the product services can be assessed through various AI-driven tools (Chatterjee et al., 2019). The application of AI and IoT significantly transformed the traditional market perceptions among users with the digital technologies' interventions by providing a better supply chain (Sujata et al., 2019). Besides brick-and-mortar stores, AI helps online businesses to develop holistic customer-centric operations and meet customer requirements (Qiu et al., 2022). Nazim Sha and Rajeswari (2019) discussed the development of AI, who also showed off an AI-supported device that can track individuals' five senses (sight, hearing, taste, smell and touch). The analytical model results showed that AI-based machine support improves consumer and product-brand associations. Artificial intelligence can support marketers in strategy and planning marketing activities by helping with segmentation, targeting, and positioning (STP), along with developing marketing strategies for future expansion (Huang & Rust, 2017; Verma et al., 2021).

AI technologies for marketing firms are proven as a good move for retaining customer bases and improving customer well-being through product-service platforms. However, many medium to small-scale firms are still dealing with digital/physical products on online platforms facing many customer retention challenges such as product visibility, product returnability, and product display over online platforms. Therefore, this chapter tries to answer the following managerial questions for organizations looking to transform their business operations through the AI application's implementation to improve their customers' well-being. Here customer well-being is discussed in the context of customer experiences, trust, ease of technology, personalization, and relationship commitment in the buyer-seller relationship. The primary research question of this chapter is to answer how relevant management theories are linked to improving customer well-being in AI-driven service quality dimensions.

The above real-world business challenging question is answered with the following objectives.

Objective 1: To identify the relevant management theories for portraying the role of AI-driven technologies in improving customer e-service quality
Objective 2: To propose the future roadmap for retail managers and customers to use AI-based technologies for a better service quality perspective

The above managerial questions are answered in this chapter with the relevant linkages among the various dimensions of customer relationships with digital tool experiences.

The rest of the chapter is arranged in the following manner. Section "Customer Relationship in an AI-Driven Market" presents the linkages between customer relations in an AI-driven market. Section "Role of Service Quality in AI-Driven Customer Services" depicts the connecting bridge between customer service quality in AI-driven experiences. The detailed dimensions of service quality and customer well-being are presented in Section "Role of Theories for Linking AI and Service Quality." Finally, the conclusion, learning, and future scope are presented in Section "AI and Service Quality Dimensions Framework."

Customer Relationship in an AI-Driven Market

The functions of customer relationship management (CRM) are widely benefited through AI user interface (AIUI) (Seranmadevi & Senthil Kumar, 2019). AIUI engages customers with service brands and enables greater experiences (Lemon & Verhoef, 2016). Many organizations respond to the customers and shape the market design with unique customer experiences that improve customer retention and connect consumers with the brand and product (Bolton et al., 2018b). In the coming years, a customer will be experiencing interaction with robots, which makes connectivity, and provides human support and a holistic man-machine system shortly (Bolton et al., 2018b; Yau et al., 2021). This is how humans will interact and transact business in the AI world. Gartner proposed a detailed strategic roadmap for developing customer experience (CX) with AI-driven technologies. Gartner's report explains the five stages of customer experience transformation for understanding user experiences.

Customer expectations and experiences are often greatly influenced by the service delivery by the retailers. Customer experiences primarily refer to a customer's overall experience with the retailer and are linked with the feelings toward the brand served by the retailer (Verhoef et al., 2009). The customer experience is based on a series of interactions with the service provider; the AI would help customers to interact on a real-time basis with the service provider (Gacanin & Wagner, 2019; Gentile et al., 2007; Huang & Rust, 2017). Therefore, many of the research contributions consider customer experiences that are conceptualized as holistic or in an individual form customer's cognitive, affective, emotional, social, and sensory elements (Lemon & Verhoef, 2016; Wamba-Taguimdje et al., 2020). Voorhees et al. (2017) emphasized that the customer experience is based on the retailer's offering, word of mouth, and organizational perspective (Kranzbühler et al., 2018), where the focus is on measuring and delivering an experience for the customer (Bolton et al., 2018b). Previous studies present a management research framework that considers the integration of autonomous machines to communicate with the customers to create knowledge with the help of big data and information, use of artificial intelligence, and then disseminate the knowledge to enhance customer relationships in a knowledge-based environment (Abu Daqar & Smoudy, 2019; Bolton et al., 2018b; Khalil & Abdelli, 2022; Yau et al., 2021). The proposed AIM framework links the three main components pre-processor, main processor, and memory storage. The main processor is the central part of the AI process, which provides the structured data and passes the primary processor to make real-time decisions and reasoning. Therefore, the AI-enabled system has characteristics that hypothesize the users' abilities, learning paradigms, and coping with human operations.

Artificial intelligence marketing (AIM) strategies ate used to automate certain marketing functions to handle a massive amount of data and process that data intelligently to support marketing mix strategies to create knowledge. Subsequently, that system is used to develop new knowledge that performs and automates marketing processes, such as generating market intelligence (Verma et al., 2021). For comparison, AIM can drill down to the individual customer level across various activities (e.g., acquisition, consumption, and disposal) related to a product or service. At the same time, traditional marketing only focuses on the firm level and acquisition/purchase activity.

Most of the past researchers consider customer experiences in four different categories.

Cognitive

This type of customer experience refers to the mental proves of the customers that link to the brand perception, memory, language, problem-solving, and abstract thinking of the individual. According to (Keiningham et al., 2017), the cognitive element of the customer experience also links to the availability of the service within the vicinity of the consumer.

Emotional

This customer experience is very complex to understand customer expectations. Ladhari et al. (2017) highlighted that emotional customer experiences are related to anger, outrage, regret, delight, joy, or surprise. All these aspects positively or negatively impact customer service quality and are linked with product quality.

Physical and Sensorial

This customer experience is often differentiated when the customers buy online and offline. For example, the offline purchase experience is linked to the features like artifacts, lighting, store layout, and signage. In contrast, online buying experiences encompass and are linked with the multiple technological advancements of the online store and the user-friendly accessibility and interfaces of the web portal (Keiningham et al., 2017).

Social Experiences

This product or brand experience is impacted by the customer's surroundings, such as friends, family, and social media. The social experience of the product also includes the social and mental identity of the customers that how they view themselves in a marketplace (Keiningham et al., 2017).

Now in the fourth industrial revolution, the question about the service or product quality and experiences is entirely different from the above four categories of customer experiences. According to a published study by Gartner, "the use of artificial intelligence-based technologies such as machine learning, natural language processing (NLP) languages understanding can help organizations to understand the customer sentiments and feedback at a different scale, precision, and speed which can is hard to achieve through the human processes." This opens a strong connection for AI-based tools for retailers to monitor and continuously improve their customer experiences, thus maintaining a competitive business environment. In many retail businesses, AI technologies are often combined with other technologies such as augmented reality, computer vision-based systems, and maintaining predictive shopping behaviors. For the above technologies to successfully improve customer service quality, it is necessary to include past customer experiences in the decision models. Leveraging AI-based tools can help monitor the customer's past buying patterns and the best way to communicate with their future buying behavior.

Role of Service Quality in AI-Driven Customer Services

Customer service quality issues are widely discussed from different perspectives. In an AI-driven business environment, customer experiences are entirely different from the experiences in another business environment (Cahyati & Seminari, 2020). Service quality research and practice are very important in the digital era when customer expectations and delivery must go hand-in-hand for improved service quality (Jain & Schultz, 2019; Narang et al., 2012). Traditionally, service quality is defined as the mismatch between the expected and perceived service when a service provider offers the requisite service through the particular brand of the product (Parasuraman et al., 1994). The service quality conception is based on the expectancy disconfirmation theory (EDC) (Collier & Bienstock, 2006). The EDC theory gives a framework for evaluating the service quality, which measures the comparison between the perception and received service based on the prior expectations of the product/service providers already offered. The EDC theoretical basis will help the researchers to link customers' e-service perspective in the digital world (Xu et al., 2020b). Service quality is broadly research domain in consumer research and discussing the various practices, outcomes, barriers, and facilitators dimensions (Cahyati & Seminari, 2020; Djakasaputra et al., 2021; Irfan et al., 2020; Jain & Schultz, 2019; Narang et al., 2012; Prentice & Kadan, 2019). Although, a lack of AI-enabled service quality and their impact on customer responses are present in the current marketing research domain. As AI-based services tend to be offered around self-service technologies (Ameen et al., 2021), the service quality differs from the interpersonal services models discussed based on the (Parasuraman et al., 1994). The AI-enabled vice quality modes considered the multiple human behavioral factors that need to be considered for improving the AI world's consumer service (Ameen et al., 2021; Gupta, 2021; Xu, Liu, et al., 2020). Ameen et al. (2021)) highlighted that managing the service quality in AI-enabled systems depends on the consumers' trust in the AI-based services. Therefore, this chapter also extends the same reason for linking the AI-driven customer service quality. The previous introduction section defines AI as machine intelligence that can be applied in various contexts to improve operational and customer well-being (Prentice et al., 2020a). AI works like humans and performs multiple cognitive tasks to achieve defined business goals and also provides a linkage to the various automated systems like computers and robots and data monitoring tools (Bolton et al., 2018a). AI-powered tools impact consumers' daily lives; for example, Apple's Siri and Google voice-assisted tools help many users connect and perform multiple tasks. The detailed application of AI at the level of concept development, operationalization, and applications are provided by (Gladson Nwokah & Ahiauzu, 2009; Kidwell et al., 2011).

AI services have received the attention of the various organizational context and add value to the product or services provided by organizations that help to improve the employees' job efficiency in terms of managing inventory, writing articles, and answering customer queries, and also enhance the customer experience through the entire customer buying cycle (Bolton et al., 2018a; Chung et al., 2020).

AI services are widely used in the hotel industry to assist guests, shape their experiences, and make them happy to stay at hotels (Li et al., 2019). For example, the Connie tool of Hilton Hotels & Resorts can translate and perform various customer-related services and help them with hotel checking (Prentice et al., 2020b). Another application of AI services is Wynn Casino in Las Vegas, where they use Amazon Echo to digitalize the hotel room services and various voice-assisted control for room lighting and temperature, in addition to offering multiple services (Prentice et al., 2020b). AI-driven services help to deploy the various robotics and their linking with human operations to improve customer experiences and facilitate location preferences and payment options (Li et al., 2019). In conclusion, AI-enabled tools and services are well transforming the services, and retail marketers need to utilize the opportunity and deploy AI-enabled services to improve the customer e-service quality. Chung et al. (2020) conducted a study to identify the relationship between the chatbot's e-services and customer toward engaging customer service experience. They used a five-dimension model for measuring the customer perception of interaction, entertainment, trendiness, customization, and problem-solving abilities of the chatbots. Thus, chatbots offer a new dimension for retailers and luxury retail brands to provide superior services for technology-savvy consumers willing to pay more for chatbot services. Further, AI-enabled chatbots ensure friendly engagement between retailers & consumers, positively impacting service quality (Chen et al., 2021; Chung et al., 2020; Haugeland et al., 2022; Sidaoui et al., 2020). The human users are well-engaged and feel comfortable with the robotic-based operations but depend on the technologies' psychological ownership and trust in post-adoption time (Delgosha & Hajiheydari, 2021).

Role of Theories for Linking AI and Service Quality

The recent work of (Ameen et al., 2021) presented a conceptual model for analyzing the role of AI tools-based consumer well-being. Researchers found that trust in digital tools is important in adopting AI-based tools for consumer well-being. Therefore, it is necessary to build and recommend a trust-based theoretical concern of the AI tool integration with consumer well-being. The trust-commitment theory (TCT) deals with the trust-measuring phenomenon in consumer well-being. TCT is a well-established theory that deals with trust commitment in a business relationship when the seller renders any product or service to the buyer (Morgan & Hunt, 1994). Trust in digital tools also depends on the seller's commitment to fulfilling the customer's needs and expectations. Therefore, integrating trust with commitment is important while measuring the relationship among any channel partner (Brown et al., 2019). The seminal work of Morgan and Hunt (1994) on trust and commitment are the main construct while measuring the relationship between trust and commitment that mediate through opportunistic behavior and communication among the seller and buyer. The application of TCT has expanded over time and is widely used to measure trust in online retailing, group buying, and online shopping behavior (Rehman et al., 2019; Wang et al., 2016). The above studies measure the consumer trust that helps the relationship between online retailers and buyers on digital platforms. To analyze the impact of digital technologies on consumer well-being (service quality), it needs to integrate the TCT with the standard service quality models or e-service quality models (Ameen et al., 2020; Goutam & Gopalakrishna, 2018).

Brown et al. (2019) developed a conceptual model for the casualty of consumer trust and commitment to identifying the key areas for improving service quality. They consider that the duration of the relationship is positively linked to the level of commitment and cost. (Gupta, 2021) discussed the role of AI-enabled customer experience in the e-commerce industry. They explored various factors such as interface design, technology, and customer service quality for measuring the trust and commitment in a seller-buyer relationship. The hypothesized mode is validated with the customers that are good users of the AI tools and claim that trust positively impacts customer service quality. The commitment in TCT-based model is considered three different forms of commitment: affective, normative, and calculative. Finally, the role of relationship commitment and its impact on AI-enabled customer experience is measured. The recent work of (Chi et al., 2021) conceptualized a formative scale for measuring consumers' trust that interacted with the AI-based social robots service delivery system. They used a third-order reflective-formative scale to measure the trust in the AI-enabled technologies for measuring the trustworthy service task and context. Many recent works consider the role of trust measurement in AI-enabled vice robots (Chi et al., 2020; Gursoy et al., 2019; Lin et al., 2020). Chi et al. (2020) presented a detailed literature review on understanding AI and its applications to service encounters and the hospitality industry that mainly discussed the role of AI technologies and their linking with the services and various adoption factors from the perspective of system design, algorithms, voice-assisted systems, and psychological knowledge. (Gursoy et al., 2019) tested an AI device use acceptance (AIDUA) model for recording the rate of consumer acceptance of the use of AI devices. The proposed model considered the three user technology acceptance levels: primary, appraisal, secondary appraisal, and outcome stage, and six technological antecedents like social influence, hedonic motivation of technology use, anthropomorphism, performance expectancy, and effort expectancy and emotion. Their model advances the TAM for accepting the multi-faced role of AI technologies in improving consumer well-being.

AI and Service Quality Dimensions Framework

The theoretical underpinning presented in Section "Role of Service Quality in AI-Driven Customer Services" is only reviewed. It considers the TCT and TAM perspectives of the new technologies (AI-enabled services). Still, a few important theories are also viewed from considering the improved level of consumer service quality and wider acceptance of AI-enabled services in various industries. Figure 11.1 presents a linkage mechanism of customer well-being in an AI-enabled world.

One of the important cognitive appraisal theories (Lazarus, 1991) can be integrated with the above models for measuring consumer trust in AI-enabled services. Second, AI-enabled services must consider the role of institutions and stakeholders in deploying digital services for consumers. Therefore, considering institutional theory (Geels, 2004; Glover et al. 2014) and multi-stakeholders theory (Freeman & McVea, 2005; Freeman & Phillips, 2005) is necessary to promote the use of AI-enabled services for improved e-service quality. Third, considering the role of level and type of intelligence is necessary for deploying AI-enabled services. Different types, such as mechanical, analytical, intuitive, and empathetic intelligence, are important for developing decision models for e-service quality (Pan et al. 2015; Porter & Heppelmann, 2014; Qiu et al., 2019). Finally, the linking of trust, familiarity with technology, social influence, robot use self-efficacy, perceived risk, and effort expectancy are some important dimensions that future exploration is required.

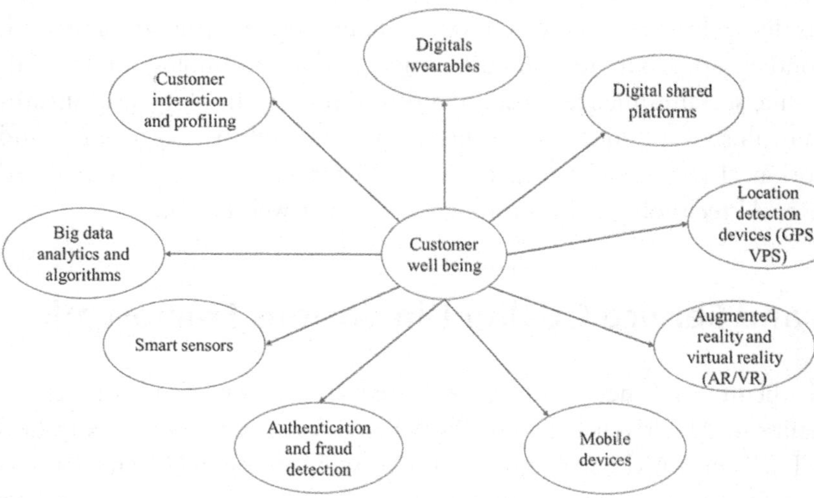

Fig. 11.1 AI technologies interventions for consumer well-being

Conclusion and Future Research Scope

With the advent of the Industry 4.0 era and the high customer expectations the service provides, it is imperative to consider customer well-being with an adequate level of digital technologies. Digital technologies considered are chatbots, AI, machine learning, NLPs, autonomous systems, and blockchain technologies. Among the above, AI and the integration of robotics for improving customer services is an emerging field for marketing researchers and organizations. The customer trust and perceived usefulness of the AI-based service system need to be robust to help consumers to connect and improve their buying behavior. This chapter presents a detailed roadmap for theoretical interventions of the management theory such as trust-commitment theory, technology acceptance model, and other relevant psychological ownership trust models for discussing the importance of AI-based tools for consumer well-being. However, the present chapter highlighted the linkage among AI technological interventions for improving consumer service quality and well-being.

This chapter helps to provide a guiding path for the market and researchers to understand AI-enabled customer experiences with improved customer service, highlighting the role of cognitive, social, emotional, and physical customer experiences and recognition of AI-enabled consumer experiences. Our work offers an early attempt to analyze how Industry 4.0 technologies (including IoT, BCT, and CPS) shape modern consumers' shopping experiences.

This chapter also presents the answer to the essential managerial questions in terms of linking management theories such as TCT, TAM, and user acceptance models to measure and evaluate consumer well-being with the advent of various AI-based tools for monitoring consumer buying behavior. The requisite service quality parameters such as perceived convenience, personalization of the services, trust in technology, service commitment, and perceived sacrifices are considered for linking the above management theories with the service quality of the consumer well-being in the AI world. The answer to the final question is to present the various psychological interventions mechanism for improving consumer well-being in an AI-driven marketplace. The chapter also presents the linkage among consumer trust, convenience factors, and personalization with the different AI-based tools, such as demand predictions, personalization of the product, a product search, and answers about how and why AI tools are essential for measuring consumer well-being.

This study is an early attempt to focus the customer experiences on AI-enabled services, which opens a new area for marketing and information systems researchers to work interdisciplinary to examine the technological interventions for improving customer well-being.

The findings of the presented chapter have some limitations in terms of the exploration and validation of the proposed construct with a real case study. It is recommended that future researchers consider the role of the presented theoretical underpinning of the management theories in improving consumer trust in AI technologies for their well-being. Therefore, future studies are recommended to consider integrating service robots with humans, where robots play an active role in improving consumer well-being in various consumer-centric markets. In addition to the above, the presented AI tools for consumer well-being in Fig. 11.1 can be modeled to obtain causal relationships among the presented variables for consumer well-being.

References

Abdulquadri, A., Kieu, T. A., & Nguyen, N. P. (2021). Digital transformation in financial services provision: Perspective to the adoption of chatbot. *Journal of Enterprising Communities: People and Places in the Global Economy, 15*(2), 258–281.

Abu Daqar, M. A. M., & Smoudy, A. K. A. (2019). International review of management and marketing the role of artificial intelligence on enhancing customer experience. *International Review of Management and Marketing, 9*(4), 22–31.

Ameen, N., Tarhini, A., Reppel, A., & Anand, A. (2021). Customer experiences in the age of artificial intelligence. *Computers in Human Behaviour, 114*, 106548. https://doi.org/10.1016/j.chb.2020.106548

Ameen, N., Tarhini, A., Shah, M., & Madichie, N. O. (2020). Going with the flow: Smart shopping malls and omnichannel retailing. *Journal of Services Marketing, 35*, 325–348. https://doi.org/10.1108/JSM-02-2020-0066

Balaji, M. S., & Roy, S. K. (2017). Value co-creation with the internet of things technology in the retail industry. *Journal of Marketing Management, 33*, 7–31. https://doi.org/10.1080/0267257X.2016.1217914

Balakrishnan, J., Nwoba, A. C., & Nguyen, N. P. (2021). Emerging-market consumers' interactions with banking chatbots. *Telematics and Informatics, 65*(5), 101711–101734.

Bolton, C., Machová, V., Kovacova, M., & Valaskova, K. (2018a). The power of human–machine collaboration: Artificial intelligence, business automation, and the smart economy. *Economics, Management, and Financial Markets, 13*, 51–56. https://doi.org/10.22381/EMFM13420184

Bolton, R. N., McColl-Kennedy, J. R., Cheung, L., Gallan, A., Orsingher, C., Witell, L., & Zaki, M. (2018b). Customer experience challenges: Bringing together digital, physical and social realms. *Journal of Service Management, 29*, 776–808. https://doi.org/10.1108/JOSM-04-2018-0113

Bowen, J., & Morosan, C. (2018). Beware hospitality industry: The robots are coming. *Worldwide Hospitality and Tourism Themes, 10*(6), 726–733. https://doi.org/10.1108/WHATT-07-2018-0045

Brown, J. R., Crosno, J. L., & Tong, P. Y. (2019). Is the theory of trust and commitment in marketing relationships incomplete? *Industrial Marketing Management, 77*, 155–169. https://doi.org/10.1016/j.indmarman.2018.10.005

Cahyati, N. K., & Seminari, N. K. (2020). The role of customer satisfaction in mediating the effect of service quality and marketing experience on repurchase intention (Study in PT Pos Indonesia expedition services). *American Journal of Humanities and Social Sciences Research, 4*(2), 128–135.

Castillo, D., Canhoto, A. I., & Said, E. (2021). The dark side of AI-powered service interactions: Exploring the process of co-destruction from the customer perspective. *Service Industries Journal, 41*, 900–925. https://doi.org/10.1080/02642069.2020.1787993

Chatterjee, S., Ghosh, S. K., Chaudhuri, R., & Nguyen, B. (2019). Are CRM systems ready for AI integration?: A conceptual framework of organisational readiness for effective AI-CRM integration. *Bottom Line, 32*(2), 144–157. https://doi.org/10.1108/BL-02-2019-0069

Chen, J. S., Le, T. T. Y., & Florence, D. (2021). Usability and responsiveness of artificial intelligence chatbot on online customer experience in e-retailing. *International Journal of Retail and Distribution Management, 49*(11), 1512–1531. https://doi.org/10.1108/IJRDM-08-2020-0312

Chi, O. H., Denton, G., & Gursoy, D. (2020). Artificially intelligent device use in service delivery: A systematic review, synthesis, and research agenda.

Journal of Hospitality Marketing and Management, 29, 757–786. https://doi.org/10.1080/19368623.2020.1721394
Chi, O. H., Jia, S., Li, Y., & Gursoy, D. (2021). Developing a formative scale to measure consumers' trust toward interaction with artificially intelligent (AI) social robots in service delivery. *Computers in Human Behavior, 118*, 106700. https://doi.org/10.1016/j.chb.2021.106700
Chung, M., Ko, E., Joung, H., & Kim, S. J. (2020). Chatbot e-service and customer satisfaction regarding luxury brands. *Journal of Business Research, 117*, 587–595. https://doi.org/10.1016/j.jbusres.2018.10.004
Collier, J. E., & Bienstock, C. C. (2006). Measuring service quality in E-retailing. *Journal of Service Research, 8*, 16. https://doi.org/10.1177/1094670505278867
Davenport, T., Guha, A., Grewal, D., & Bressgott, T. (2020). How artificial intelligence will change the future of marketing. *Journal of the Academy of Marketing Science, 48*, 24–42. https://doi.org/10.1007/s11747-019-00696-0
Delgosha, M. S., & Hajiheydari, N. (2021). How human users engage with consumer robots? A dual model of psychological ownership and trust to explain post-adoption behaviours. *Computers in Human Behavior, 117*, 106660. https://doi.org/10.1016/j.chb.2020.106660
Djakasaputra, A., Wijaya, O. Y. A., Utama, A. S., Yohana, C., Romadhoni, B., & Fahlevi, M. (2021). Empirical study of indonesian SMEs sales performance in digital era: The role of quality service and digital marketing. *International Journal of Data and Network Science, 5*(3), 163–494. https://doi.org/10.5267/j.ijdns.2021.6.003
Freeman, R. E. E., & McVea, J. (2005). A stakeholder approach to strategic management. *SSRN Electronic Journal, 2005*, 32. https://doi.org/10.2139/ssrn.263511
Freeman, R. E. E., & Phillips, R. A. (2005). Stakeholder theory: A libertarian Defense. *SSRN Electronic Journal, 2005*(21). https://doi.org/10.2139/ssrn.263514
Gacanin, H., & Wagner, M. (2019). Artificial intelligence paradigm for customer experience Management in Next-Generation Networks: Challenges and perspectives. *IEEE Network, 33*(2), 188–194. https://doi.org/10.1109/MNET.2019.1800015
Geels, F. W. (2004). From sectoral systems of innovation to socio-technical systems: Insights about dynamics and change from sociology and institutional theory. *Research Policy, 33*, 897–920. https://doi.org/10.1016/j.respol.2004.01.015

Gentile, C., Spiller, N., & Noci, G. (2007). How to sustain the customer experience: An overview of experience components that co-create value with the customer. *European Management Journal, 25*, 395–410. https://doi.org/10.1016/j.emj.2007.08.005

Gladson Nwokah, N., & Ahiauzu, A. I. (2009). Emotional intelligence and marketing effectiveness. *Marketing Intelligence & Planning, 27*, 864–881. https://doi.org/10.1108/02634500911000199

Glover, J. L., Champion, D., Daniels, K. J., & Dainty, A. J. D. (2014). An institutional theory perspective on sustainable practices across the dairy supply chain. *International Journal of Production Economics, 152*, 102–111. https://doi.org/10.1016/j.ijpe.2013.12.027

Goutam, D., & Gopalakrishna, B. V. (2018). Customer loyalty development in online shopping: An integration of e-service quality model and commitment-trust theory. *Management Science Letters, 8*, 1149–1158. https://doi.org/10.5267/j.msl.2018.8.009

Gupta, A. (2021). Exploratory analysis of factors influencing Ai-enabled customer experience for E-Commerce industry. *Bioscience Biotechnology Research Communications, 14*, 104–112. https://doi.org/10.21786/bbrc/14.5/21

Gursoy, D., Chi, O. H., Lu, L., & Nunkoo, R. (2019). Consumers acceptance of artificially intelligent (AI) device use in service delivery. *International Journal of Information Management, 49*, 157–169. https://doi.org/10.1016/j.ijinfomgt.2019.03.008

Haugeland, I. K. F., Følstad, A., Taylor, C., & Alexander, C. (2022). Understanding the user experience of customer service chatbots: An experimental study of chatbot interaction design. *International Journal of Human Computer Studies, 161*, 102788. https://doi.org/10.1016/j.ijhcs.2022.102788

Huang, M. H., & Rust, R. T. (2017). Technology-driven service strategy. *Journal of the Academy of Marketing Science, 45*, 906–924. https://doi.org/10.1007/s11747-017-0545-6

Huang, M. H., & Rust, R. T. (2021). Engaged to a robot? The role of AI in service. *Journal of Service Research, 24*. https://doi.org/10.1177/1094670520902266

Irfan, A., Mahfudnurnajamuddin, M., Hasan, S., & Mapparenta, M. (2020). The effect of destination image, service quality, and marketing mix on tourist satisfaction and revisiting decisions at tourism objects. *International Journal of Multicultural and Multireligious Understanding, 7*, 727–740. https://doi.org/10.18415/ijmmu.v7i8.2046

Jain, V., & Schultz, D. E. (2019). How digital platforms influence luxury purchase behaviour in India? *Journal of Marketing Communications, 25*, 41–64. https://doi.org/10.1080/13527266.2016.1197295

Keiningham, T., Ball, J., Benoit (née Moeller), S., Bruce, H. L., Buoye, A., Dzenkovska, J., Nasr, L., Ou, Y. C., & Zaki, M. (2017). The interplay of customer experience and commitment. *Journal of Services Marketing, 31*, 148–160. https://doi.org/10.1108/JSM-09-2016-0337

Khalil, A., & Abdelli, M. E. A. (2022). Do digital technologies influence the relationship between the COVID-19 crisis and SMEs' resilience in developing countries? *Journal of Open Innovation: Technology, Market, and Complexity, 8*(2), 100–109.

Kidwell, B., Hardesty, D. M., Murtha, B. R., & Sheng, S. (2011). Emotional intelligence in marketing exchanges. *Journal of Marketing, 75*, 78–95. https://doi.org/10.1509/jmkg.75.1.78

Kranzbühler, A. M., Kleijnen, M. H. P., Morgan, R. E., & Teerling, M. (2018). The multilevel nature of customer experience research: An integrative review and research agenda. *International Journal of Management Reviews, 20*, 433–456. https://doi.org/10.1111/ijmr.12140

Ladhari, R., Souiden, N., & Dufour, B. (2017). The role of emotions in utilitarian service settings: The effects of emotional satisfaction on product perception and behavioural intentions. *Journal of Retailing and Consumer Services, 34*, 10–18. https://doi.org/10.1016/j.jretconser.2016.09.005

Lazarus, R. S. (1991). Progress on a cognitive-motivational-relational theory of emotion. *American Psychologist, 46*, 819–834. https://doi.org/10.1037/0003-066X.46.8.819

Lemon, K. N., & Verhoef, P. C. (2016). Understanding customer experience throughout the customer journey. *Journal of Marketing, 80*. https://doi.org/10.1509/jm.15.0420

Li, J. (. J.)., Bonn, M. A., & Ye, B. H. (2019). Hotel employee's artificial intelligence and robotics awareness and its impact on turnover intention: The moderating roles of perceived organisational support and competitive psychological climate. *Tourism Management, 73*. https://doi.org/10.1016/j.tourman.2019.02.006

Li, M., Yin, D., Qiu, H., & Bai, B. (2021). A systematic review of AI technology-based service encounters: Implications for hospitality and tourism operations. *International Journal of Hospitality Management, 95*, 102930. https://doi.org/10.1016/j.ijhm.2021.102930

Lin, H., Chi, O. H., & Gursoy, D. (2020). Antecedents of customers' acceptance of artificially intelligent robotic device use in hospitality services. *Journal of Hospitality Marketing and Management, 29*, 530–549. https://doi.org/10.1080/19368623.2020.1685053

Mogaji, E., & Nguyen, P. N. (2022). Managers' understanding of artificial intelligence in relation to marketing financial services: Insights from a cross-country study. *International Journal of Bank Marketing, 40*(6), 1272–1298.

Mogaji, E., Soetan, T., & Kieu, T. (2020). The implications of artificial intelligence on the digital marketing of financial services to vulnerable customers. *Australasian Marketing Journal., 29*(3), 235–242.

Morgan, R. M., & Hunt, S. D. (1994). The commitment-trust theory of relationship marketing. *Journal of Marketing, 58*, 20–38. https://doi.org/10.1177/002224299405800302

Narang, S., Jain, V., & Roy, S. (2012). Effect of QR codes on consumer attitudes. *International Journal of Mobile Marketing, 7*, 52–54.

Nazim Sha, S., & Rajeswari, M. (2019). Creating a brand value and consumer satisfaction in E-commerce business using artificial intelligence with the help of vosag technology. *International Journal of Innovative Technology and Exploring Engineering, 8*, 2. https://doi.org/10.1177/002224299405800302

Pan, Y., Okada, H., Uchiyama, T., & Suzuki, K. (2015). On the reaction to Robot's speech in a hotel public space. *International Journal of Social Robotics, 7*, 911–920. https://doi.org/10.1007/s12369-015-0320-0

Parasuraman, A., Zeithaml, V. A., & Berry, L. L. (1994). Reassessment of expectations as a comparison standard in measuring service quality: Implications for further research. *Journal of Marketing, 58*, 111–124. https://doi.org/10.1177/002224299405800109

Porter, M. E., & Heppelmann, J. E. (2014). How smart, connected products are transforming competition. *Harvard Business Review*. https://doi.org/10.1177/002224299405800302

Prentice, C., Dominique Lopes, S., & Wang, X. (2020a). Emotional intelligence or artificial intelligence– An employee perspective. *Journal of Hospitality Marketing and Management, 29*, 377–403. https://doi.org/10.1080/19368623.2019.1647124

Prentice, C., Dominique Lopes, S., & Wang, X. (2020b). The impact of artificial intelligence and employee service quality on customer satisfaction and loyalty. *Journal of Hospitality Marketing and Management, 29*, 739–756. https://doi.org/10.1080/19368623.2020.1722304

Prentice, C., & Kadan, M. (2019). The role of airport service quality in airport and destination choice. *Journal of Retailing and Consumer Services, 47*, 40–48. https://doi.org/10.1016/j.jretconser.2018.10.006

Qiu, H., Li, M., Bai, B., Wang, N., & Li, Y. (2022). The impact of AI-enabled service attributes on service hospitableness: The role of employee physical and psychological workload. *International Journal of Contemporary Hospitality Management, 34*, 1374–1398. https://doi.org/10.1108/IJCHM-08-2021-0960

Qiu, H., Li, M., Shu, B., & Bai, B. (2019). Enhancing hospitality experience with service robots: The mediating role of rapport building. *Journal of Hospitality Marketing and Management, 29*, 247–268. https://doi.org/10.1080/19368623.2019.1645073

Rehman, S. U., Bhatti, A., Mohamed, R., & Ayoup, H. (2019). The moderating role of trust and commitment between consumer purchase intention and online shopping behaviour in the context of Pakistan. *Journal of Global Entrepreneurship Research, 9*, 43. https://doi.org/10.1186/s40497-019-0166-2

Rouhani, S., Ashrafi, A., Zare Ravasan, A., & Afshari, S. (2016). The impact model of business intelligence on decision support and organizational benefits. *Journal of Enterprise Information Management, 29*, 19–50. https://doi.org/10.1108/JEIM-12-2014-0126

Russel, S., & Norvig, P. (2012). Artificial intelligence—A modern approach 3rd edition. *The Knowledge Engineering Review, 1*, 78–79. https://doi.org/10.1017/S0269888900007724

Seranmadevi, R., & Senthil Kumar, A. (2019). Experiencing the AI emergence in Indian retail – Early adopters approach. *Management Science Letters, 9*, 33–42. https://doi.org/10.5267/j.msl.2018.11.002

Sidaoui, K., Jaakkola, M., & Burton, J. (2020). AI feel you: Customer experience assessment via chatbot interviews. *Journal of Service Management, 31*, 745–766. https://doi.org/10.1108/JOSM-11-2019-0341

Sujata, M., Khor, K. S., Ramayah, T., & Teoh, A. P. (2019). The role of social media on recycling behaviour. *Sustainable Production and Consumption, 20*, 365–374. https://doi.org/10.1016/j.spc.2019.08.005.

Verhoef, P. C., Lemon, K. N., Parasuraman, A., Roggeveen, A., Tsiros, M., & Schlesinger, L. A. (2009). Customer experience creation: Determinants, dynamics and management strategies. *Journal of Retailing, 85*, 31–41. https://doi.org/10.1016/j.jretai.2008.11.001

Verma, S., Sharma, R., Deb, S., & Maitra, D. (2021). Artificial intelligence in marketing: Systematic review and future research direction. *International*

Journal of Information Management Data Insights, 1, 100002. https://doi.org/10.1016/j.jjimei.2020.100002

Voorhees, C. M., Fombelle, P. W., Gregoire, Y., Bone, S., Gustafsson, A., Sousa, R., & Walkowiak, T. (2017). Service encounters, experiences and the customer journey: Defining the field and a call to expand our lens. *Journal of Business Research, 79*, 269–280. https://doi.org/10.1016/j.jbusres.2017.04.014

Wamba-Taguimdje, S. L., Fosso Wamba, S., Kala Kamdjoug, J. R., & Tchatchouang Wanko, C. E. (2020). Influence of artificial intelligence (AI) on firm performance: The business value of AI-based transformation projects. *Business Process Management Journal, 26*, 1893–1924. https://doi.org/10.1108/BPMJ-10-2019-0411

Wang, J., & Fan, X. (2020). Co-production strategy, retail competition, and market segmentation. *Asia Pacific Journal of Marketing and Logistics, 32*, 607–630. https://doi.org/10.1108/APJML-10-2018-0408

Wang, W. T., Wang, Y. S., & Liu, E. R. (2016). The stickiness intention of group-buying websites: The integration of the commitment–trust theory and e-commerce success model. *Information and Management, 53*, 625–642. https://doi.org/10.1016/j.im.2016.01.006

Xu, Y., Shieh, C. H., van Esch, P., & Ling, I. L. (2020a). AI customer service: Task complexity, problem-solving ability, and usage intention. *Australasian Marketing Journal, 28*, 189–199. https://doi.org/10.1016/j.ausmj.2020.03.005

Xu, Z., Liu, W., Huang, J., Yang, C., Lu, J., & Tan, H. (2020b). Artificial intelligence for securing IoT services in edge computing: A survey. *Security and Communication Networks, 2020*. https://doi.org/10.1155/2020/8872586

Yau, K. L. A., Saad, N. M., & Chong, Y. W. (2021). Artificial intelligence marketing (aim) for enhancing customer relationships. *Applied Sciences (Switzerland)*. https://doi.org/10.3390/app11188562

12

Role of Risk on Employees' Willingness to Collaborate with Artificial Intelligence and Its Impact on Wellbeing

Stefanie Paluch and Thomas Wittkop

Introduction

Technology is omnipresent in the workplace. Year after year capabilities of technologies increases steadily, paving the way for self-reliantly working technology (Huang & Rust, 2018; Schepers et al., 2022). This capability growth turns workplace technology from passive devices (e.g., laptops, smartphones) into intelligent co-working service technologies (e.g., service robots, intelligent forecasting algorithms, chatbots), which can augment or substitute their human colleagues by fulfilling services within the company and for the customer.

S. Paluch (✉)
RWTH Aachen University, Aachen, Germany
e-mail: paluch@time.rwth-aachen.de

T. Wittkop
OWL University of Applied Sciences and Arts, Lemgo, Germany
e-mail: thomas.wittkop@th-owl.de

© The Author(s), under exclusive license to Springer Nature Switzerland AG 2023

J. N. Sheth et al. (eds.), *Artificial Intelligence in Customer Service*,
https://doi.org/10.1007/978-3-031-33898-4_12

Despite the impressive technological advances, implementing and getting employees to collaborate with innovative technology poses a great challenge (Paluch et al., 2022). On the one hand, employees benefit from improvements made possible by intelligent technology in essential business areas, such as the capacity to interact with customers and store and analyze customer data (Rust & Huang, 2014). However, intelligent technology may also be viewed negatively by workers as a threat and a competitor (Jörling et al., 2019) and possibly influences their wellbeing (Makridis & Mishra, 2022). Hence, it is proposed that employees' perceived risk of co-working service technologies is the main barrier that hinders their willingness to collaborate with these technologies.

Previous research has mainly focused on the external customer perspective but needs to consider the internal employee perspective. However, this is of utmost importance since employees' collaboration with technology is key to a company's success (McLeay et al., 2021; Paluch et al., 2022). Moreover, Huang and Rust (2018) underlined the necessity to investigate and actively define conditions for the co-existence of intelligent technologies and human employees. To fill this gap, this research aims to identify factors influencing the employees' perception of co-working service technologies, which affect employees' willingness to collaborate with intelligent technology in the working environment and ultimately impact employees' wellbeing. This chapter makes three contributions. First, the article theoretically establishes the relevance and relation between employees' perceived wellbeing of co-working service technologies on employees' willingness to collaborate with the technology. Second, an explorative interview study with 36 participants is conducted, and dimensions such as technology characteristics, job demands, and resources are identified, directly influencing employees' wellbeing. Third, based on qualitative research and the literature analysis, constructs and their relations with each other are discussed, and the conceptual research model on employees' willingness to collaborate is derived. Thus, the literature on innovative service technologies (Jörling et al., 2019; Mende et al., 2019; van Doorn, 2017; Wirtz et al., 2018), technologies' affection on jobs (Huang & Rust, 2018), and frontline employees (McLeay et al., 2021) is linked with the literature from the field of collaboration and teamwork (Birdi et al., 2008; Boos et al., 2022; Jones & George, 1998).

Conceptual Background

Artificial Intelligence and Services

This article focuses on artificial intelligence–based services (AIBS or AI-based services), which are defined as *information technology in a virtual (e.g., chatbot) or physical (e.g., service robot) embodiment, providing services in a company to humans or objects by performing both virtual (e.g., data analysis, predictions, and decision-making) and physical tasks (e.g., cleaning, manufacturing, serving) with a higher degree of autonomy* (Huang & Rust, 2018; Jörling et al., 2019; Kunz et al., 2018; Wirtz et al., 2018). Due to technical advances, AI-based technologies can be integrated into various industries (e.g., healthcare, finance, hospitality), giving companies entirely new options for designing internal structures, processes, and customer service. For example, service robots appear in large hotel chains worldwide and welcome guests, check them in, or help them make restaurant reservations. Other examples are avatars used at the arts to help passengers find their gates or show them the direction to the connection terminal. In addition, there is research on how service situation robots can be better than human employees (Pitardi et al., 2021) and how processes should be designed so that human employees and AI-based services can successfully work together in hybrid teams (McLeay et al., 2021; Mogaji & Nguyen, 2022).

Related research streams emphasize the relevance of investigating collaboration with intelligent technology (You & Robert, 2018). They state: "If professional service robots are to share the workplace with people, we need to understand what their interaction is likely to be like. [...] The better we understand these questions, the better we can design robots to be effective work partners."

Prior research indicated that teams performed better when teammates were emotionally attached to the technology (You & Robert, 2018). In contrast, other scholars propose focusing less on the emotional perspective and establishing a more machine-like appearance of technology (Hinds et al., 2004). Besides those scattered findings, little is known about the best way to create service technology to co-exist with employees in the workplace (You & Robert, 2018).

Willingness to Collaborate with Technology

The main consideration is collaboration within an organization because it is essential to the success of the business. Collaboration can be defined as "socially contrived mechanisms for collective action, which are continually shaped and restructured by actions and symbolic interpretations of the parties involved" (Ring & van de Ven, 1994, p. 96). In this definition, social perspective is highlighted, which generally influences individuals' willingness to collaborate (Smith et al., 1995). Social factors are, for example, individuals' relations, personal characteristics, and trust that individuals perceive in each other (Smith et al., 1995).

Even though technology can evoke a feeling of social presence (van Doorn, 2017) and computers are perceived as social actors (Nass et al., 1994), individuals' willingness to collaborate with human beings differs from the willingness to collaborate with technology. For example, the willingness to use advice from an imperfect technology is significantly lower than that of a human being who has previously failed (Dietvorst et al., 2018; Prahl & van Swol, 2017).

Empirical Study: Exploring Employees' Willingness to Collaborate with AI and Its Impact on Wellbeing

Qualitative Studies

Qualitative research was chosen as the methodological foundation for this study. The collaboration between AI-based services and human employees is a fairly new phenomenon, and empirical research on this topic still needs to be improved. Therefore, an exploratory research approach was chosen to set up a holistic framework of conditional factors influencing employees' willingness to collaborate with artificial intelligence in service contexts. The data collection consisted of qualitative problem-centered interviews and a focus group discussion to gather insights on individual employees' perceptions of technology infusion in the workplace.

Selection of Interview Partners

Considering the assumption that the influence of artificial intelligence in the working environment will continue to increase across industries in the future, a high percentage of professions will be directly or indirectly affected. Therefore, the sample was heterogeneous, with a maximum variation of interview partners and their professional backgrounds (Patton, 2015). In sum, 31 interview partners (19 female; 12 male) were selected for the personal interviews. In addition, five participants (two female; three male) were chosen for the focus group discussion as it is said to be the optimal number for focus group discussions (Krueger, 2014). The sample covered sectors and professions from logistics, electrical engineering, asset management, automotive, human resources, personnel consulting, chemistry, coaching, student counseling, tourism, marketing, project management, and sales. The ages ranged from 24 to 61 years. The interview partners were contacted through personal contacts and the professional network LinkedIn. The interviews lasted between 39 and 91 minutes and were audio-recorded with the permission of the interview partners.

Semi-structured Interview Guideline

The interview guideline is an important instrument for problem-centered interviews since it provides structure and orientation for the interviewer. It also supports the comparability of the interviews. The interview guideline developed for this study consists of open questions within four main sections. In the first part of the interview, questions focused on general knowledge and experiences with AI-based services in everyday life. In the second part, questions were related to the general knowledge, requirements, and influence of AI on the working environment and individual attitudes toward the topic. In the third part, the questions were concretized to the interviewee's ideas about their collaboration with AI. Finally, the last section covered questions about the future developments and scenarios of AI–human collaboration.

Data Analysis

The interview material is systematically structured and coded using the qualitative software NVivo (Bazeley & Jackson, 2013). Content analysis is applied to categorize the data material. Sentence-by-sentence, the data material is analyzed to identify common themes, patterns, and categories emerging from the data, following an inductive coding strategy (Patton, 2015). Within NVivo, the 31 transcripts and the transcript from the focus group were processed one after the other, whereby the individual sentences were checked, relevant parts marked, and then assigned to one or more codes. Though, the entire coding process aims to successively refine and optimize the overall research question (Patton, 2015). This process will continue until no further categories emerge and a theoretical saturation is reached (Strauss & Corbin, 1998). In total, three main dimensions—(1) technology characteristics, (2) job characteristics, and (3) employee characteristics—and several subcategories were identified during the analysis that captures the facilitating conditions of employees' willingness to collaborate with AI-based services in the workplace. The conceptual research model is set up based on the results from the qualitative interview study and synchronized with insights from existing literature.

Conceptualizing a Framework for Employees' Willingness to Collaborate with AI-Based Services and Its Impact on Wellbeing

Based on the findings in our qualitative study and insights from the literature, the framework is derived. The results from the qualitative study were merged into more abstract constructs. Therefore, each research team member went through all the identified dimensions independently and ranked their relevance to the supposed impact the dimension has on employees' risk perception of collaborating technology. Afterward, the team discussed the individual rankings and agreed on a final set of constructs. The constructs are (1) technology characteristics, covering the level of automation and social presence; (2) job characteristics, including demands and resources; (3) employee characteristics covering employees' personality, self-efficacy, and voluntary use; and (4) control. On that basis, the framework is proposed, described, explained, and enriched with quotations from the qualitative interviews in the following section (Fig. 12.1).

This framework proposes risk as a central conditional factor that influences employees' willingness to collaborate with AI-based services and ultimately impacts employees' wellbeing. Humans generally perceive new technologies as risky (Featherman & Pavlou, 2003) since they expect possible negative outcomes when using them. A reason for this might be the lack of transparency, which is related to the lack of expertise. Employees do not fully understand how AI works in the early implementation phase. Thus, interview partners expect support from their employers. They insist on receiving proper training and instructions, allowing them to become more familiar with the topic and understand AI in greater detail. *More training and information so I know how this works* [IP22]. They aim to fully understand how decisions were made in the past and how AI will

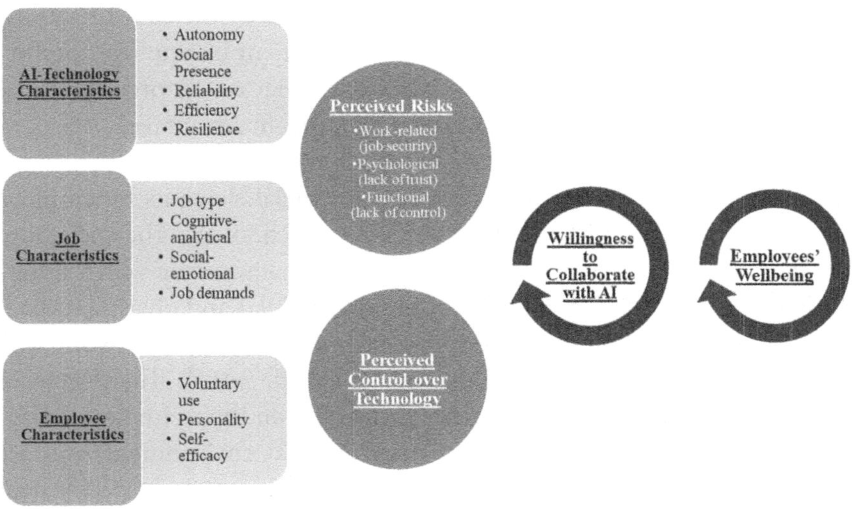

Fig. 12.1 Model of the employees' wellbeing when collaborating with AI (own illustration)

make future decisions. Another more abstract risk facet is the risk of occupational changes, which can be attributed to employees' fear of personal and professional consequences. These statements relate to the general perception of labor market changes and underline employees' fear of losing their jobs (Fraune et al., 2020) and other related professional consequences that might occur when using AI in the workplace. *I'm scared if the robots can do my job, then I'm obsolete. Why do we need workers, then?* [IP20]. In the following, the main dimensions of the framework are presented and discussed.

Technology Characteristics

The first dimension summarizes technology-related requirements for a successful collaboration with AI and challenges impacting employees' engagement. All interview partners generally expect the technology to be useful in their workplace and easy to use, ideally so that employees can use it intuitively. In addition to these basic requirements, they expect reliability and functionality so that errors and failures do not occur or can be reduced.

AI is anticipated to build greater and more consistent resilience than human workers, increasing reliability. The increased efficiency of various workflows, which has led to an overall improvement in employee performance on an individual level and work outcomes on an organizational level, is another area where employees recognize an advantage.

Based on further insights from the qualitative study, the technology's degree of autonomy and social presence are crucial elements for employees' willingness to collaborate with the service technology. The autonomy of technology is probably the most important characteristic which differentiates the technology from other common products and physical devices (Thrun, 2004). Autonomy describes a technology's ability to work self-reliantly and without human interference (Wolf, 2015). The degree of autonomy that technology has varies depending on the technologies ability to "sense its environment, plan based on that environment, and act upon that environment with the intent of reaching some task-specific goal (either given to or created by the robot) without external control" (Beer et al., 2014, p. 77), often described on a continuum, ranging from no autonomy to full autonomy. The highest level of autonomy describes a technology, which "decides everything, acts autonomously, ignores the human" (Sheridan & Verplank, 1978, pp. 168–170) and might therefore be considered as risky and uncontrollable by employees.

Focusing on our research purpose and potential barriers to employees' willingness to collaborate, interview partners mentioned several concerns directly related to the autonomy of AI-based services, for example, security issues, a lack of control and higher risk perception that prevailed during the interviews. For example, employees might feel observed by the integrated cameras, or their performance is constantly monitored, and worry about data security due to the connection to the internet (Wirtz et al., 2022). In addition, a major topic for employees is trust, which in their opinion is hard to establish with AI because many of them do not have the technology expertise and cannot build on former experience. *Exactly, I have to see what AI can do and understand how it works because I don't think that otherwise there would be a basis of trust somehow. I'm skeptical about the matter, [...] so trust is such a point that definitely has to be there* [IP5].

The second technology dimension of AI-based services is social presence. Social presence describes the "sense of being with another" (van Doorn, 2017). Traditionally, social presence refers to face-to-face communication between two human beings. But, with increasing technology intelligence and more sophisticated interactions between technologies and humans, customers may perceive a social and emotional connection with the technology (Lin et al., 2021; van Doorn, 2017). However, it is currently difficult for workers to envisage a relationship, and the interview partners believe that social presence can rarely be delivered by AI. *I don't think it's pleasant to just work with machines all day. Small talk with colleagues is also part of the job* [IP19]. Hence, this traditional perspective has shifted in the context of technological evolution to consider "quasi-social relationships" (Biocca & Harms, 2002) between humans and technology as highly relevant for service research since it affects customers' service experience and service outcomes (Jörling et al., 2019; Mende et al., 2019; van Doorn, 2017). Even though social presence is influenced by the technology's autonomy (Breazeal, 2003), social presence is considered an additional construct for the technology dimension. This has two reasons. First, social presence is also affected by many other factors (e.g., technology's embodiment), and second, it has strong relevance to people's handling and management of technologies (Jörling et al., 2019).

Job Characteristics

Employees' willingness to collaborate with AI-based services also depends on the type of job. Technology will impact different jobs differently (Wirtz et al., 2018), and employees' roles will change over time (Larivière et al., 2017). Building on the Service Robot Deployment (SDR) model from Wirtz et al. (2018), job tasks are classified based on the level of cognitive/analytical and social/emotional skills. This model predicts robots will deliver some jobs, others by human employees and human–robot teams will do the third kind of job (Fig. 12.2).

The underlying assumption of the SDR model is that AI better handles complex decision-making situations in which cognitive/analytical skills are highly demanded. In contrast to the theoretical considerations, the interview partners do not believe AI can fulfill complex tasks. They

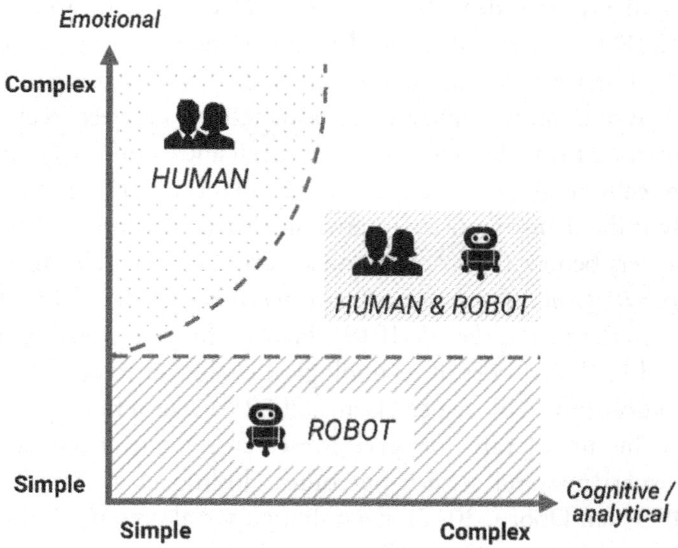

Fig. 12.2 Service Robot Deployment (SDR) model (Wirtz et al., 2018)

assume that activities taken over by AI were mainly monotonous, standard repetitive tasks. *Um, I see advantages in the fact that technology can perform very repetitive, blunt tasks that an employee should not take over; I rather see the potential for AI there* [IP10]. A further area in which AI was considered beneficial is physically strenuous activities. By transferring such tasks to technology, employees' health is protected, downtimes of the workforce can be reduced, and their health improved.

Humans can show real emotions since they intuitively react to certain situations and are better at displaying social/emotional skills.

In this respect, the employees were assigned activities requiring their complex thinking and creativity. *For me, it is effective if you leave the routine activities to artificial intelligence, and everything that needs a human approach, or in some way creativity if you use the worker there* [IP29].

The respondents have very mixed views regarding customer contact, and their opinions do not coincide. Some see human employees assigned to direct customer contact because interacting with human employees gives customers a good feeling, and they feel better supported. But on the other hand, others said it is only a matter of time until employees and customers get used to artificial contact; as soon as AI is established in the service environment, new areas of responsibility can arise for the workforce that lies in the supervision and monitoring of the AI *[…] there will certainly be a lot of changes, there may be new professions, which then just deals with the monitoring of the AI, with the repair, with the design of the AI, […], but also simply then professions may also disappear somehow then, the really monotonous performing activities, assembly line, maybe a gatekeeper […].* [IP 14].

As advances in artificial intelligence inevitably disrupt labor markets, the dynamics within any given job are subject to change (Frank et al., 2019). While some service technologies are making entire positions obsolete, others can change professions in a dissatisfactory way for employees. Ultimately, changes to the job lead to changes in willingness to cooperate. To characterize and model the impact of AI-based services on jobs, the job demands-resources model is a well-established concept describing any job in its different facets (Demerouti et al., 2001). Job demands can be described as "things to be done" associated with physical or psychological effort, which requires physical or psychological costs (Schaufeli & Bakker,

2004). Job resources describe the physical, psychological, social, or organizational characteristics of a job that (1) reduce job demands and psychological or physical cost, (2) aid in work goal achievement, and (3) lead to personal growth, learning, and development (Demerouti et al., 2001). These proposed resources are located at the organizational level (e.g., salary, job security), social level (e.g., supervisor support, team climate), work organization (e.g., roles, participation in decision-making), and task level (e.g., skill variety, task significance) (Bakker et al., 2004). Increases in job demands typically lead to exhaustion, while a lack of job resources is associated with disengagement (Hakanen et al., 2008). Exhaustion has been shown to increase the risk of burnout, while disengagement prevents a positive and fulfilling state of mind that describes vigor, dedication, and absorption (Schaufeli & Bakker, 2004).

Employee Characteristics

Individual employee-related characteristics must be considered to foster a willingness to collaborate with AI-based services. Interview partners mentioned that collaboration with other colleagues and different characters could lead to multiple advantages, such as knowledge sharing and discovering innovation possibilities. If colleagues were completely replaced, this opportunity would disappear. In general, a basic condition for successful collaboration is voluntary usage. Interview partners demand a self-determined application for which tasks and service contexts they collaborate with AI. For this purpose, the choice between several technologies or different levels of automation should be given. A fundamental, unbiased approach of employees toward AI is beneficial to facilitate the subsequent implementation steps. In addition, some employees prefer to complete tasks alone rather than in a team, so assistance is refused. It also depends on the *employee's personality, how important the job is for the individual, and* how far the employee identifies with the job. Employees in higher positions do not see their job in danger because they consider themselves and their roles very important to the company. *Well, I think it's hurting your self-confidence when you must let a machine tell you what to do and what not to do* [IP6]. Also, the degree of employees' self-efficacy might influence the willingness to collaborate.

The Dilemma Between Risk and Control

A central element in the proposed framework control, which can be defined as a person's mastery over the environment (White, 1959), also covers technology in the working environment. Control is the belief that "events are influenced by and contingent upon one's behavior and not fate, circumstances, other people, or uncontrollable physical forces" (Inesi et al., 2011, p. 1042). Control represents a universal human preference, need, and driving force, motivating people to restore control in case it is lost (Inesi et al., 2011). In service research, it has been demonstrated that control increases the probability of using self-service technology, being more satisfied with it, and perceiving more value in the technology (Bateson, 1985; Collier & Sherrell, 2010). Recent research has additionally shown that control is important when consumers use service robots (Jörling et al., 2019). Also, our interview partners mentioned control as a possibility to overcome the higher risk perception. It just depends. *If I can control the technology,* i.e., *can switch it on and off at any time, when I want to, then that's ok. I just have to be able to control everything* [IP33].

The role of control and technology usage. It has been demonstrated that the design and form of technologies impact customers' perception of control over the technology (Jörling et al., 2019). Furthermore, with increasing technology autonomy, the user is replaced as an active-decision maker (Wolf, 2015), decreasing the user's interaction and direct interference with the technology and, finally, the control over the technology (Huang & Rust, 2018).

Besides the degree of autonomy, a technology's social presence directly influences the control users perceive over the technology. For thousands of years, human interaction exclusively included face-to-face communication with other human beings, shaping the human brain and behavior (Darwin, 1859; Kock, 2004). Hence, control was generally only relevant and present in human social interactions. However, humans tend to apply and transfer the mechanisms they learned in human–human interactions to human–technology interactions (Rahwan et al., 2019). Moreover, human beings even transfer social and humanoid cues to non-human entities (e.g., clouds) to understand their behavior and have an increased feeling of control (Epley et al., 2007). Therefore, we propose a relationship between social presence and customers' perception of control over the technology.

The role of risk and technology usage. The impact of risk perception on adopting technology is well-established in the literature (Pavlou, 2003). For example, Li et al. point out that "users must overcome perceptions of risk and uncertainty before using a novel technology" (2008, p. 39). However, the risk does not only influence the adoption of new technology, but the perception of risk is, in turn, influenced by the characteristics of the technology (Montoya-Weiss et al., 2003). As described above, this study points out a high degree of autonomy as the main characteristic of AI-based services. Since "technologically sophisticated products generally lead consumers to perceive more risk" (Rijsdijk & Hultink, 2009, p. 29) and an increased product autonomy increases the perceived risk (Rijsdijk & Hultink, 2003), this research argues that AI-based services are also perceived as risky.

Besides the high degree of autonomy, the social presence of an AI-based service is considered essential for employees' willingness to collaborate with the technology. However, the impact of this construct might be ambiguous. On the one hand, increasing social presence may increase the risk perception for people who feel insecure in social interactions (Kim & McGill, 2011). On the other hand, lowering the social presence may increase the perception of risk since intangibility creates a high-risk perception for service consumers (Paluch & Wunderlich, 2016). Hence, it is argued that social presence impacts risk perception, but the direction depends on further external circumstances.

The dilemma between risk and control. Risk-taking behavior is influenced by perceived control, which has been demonstrated in different areas; for example, food safety (Frewer et al., 1994), health risks (Weinstein, 1980), sexual behavior (Klein & Kunda, 1994), driving behavior (McKenna, 1993), and gambling games (Chau & Phillips, 1995). In a technological context, a perceived lack of control is the main driver for negative emotions toward the technology Rijsdijk & Hultink, 2003). Control is a universal human preference (Langer & Rodin, 1976). Surprisingly, people prefer riskier situations, in which they perceive control, to less risky situations, in which people do not perceive control (Klein & Kunda, 1994). However, if people perceive control over something, they perceive the situation as less risky (Horswill & McKenna, 1999). This relation is even valid when people only have an illusion of control. For example, when perceived control was increased in a gambling context, study participants also increased their bet sizes because they perceived that positive outcomes resulted from their skill even though it was just a coincidence (Chau & Phillips, 1995). This relation also holds (potentially) risk-bearing entities (e.g. AI-based services). Kim and McGill (2011) have demonstrated that if people think they can control outcomes from a risk-bearing entity, they perceive it as less risky. Hence, a strong relation between control and risk is proposed: the higher perceived control over the AI-based services, the less it is perceived as risky.

General Discussion and Implications

Due to their (future) superior job-performing abilities and increasing computational capabilities, working with AI-based services will be inevitable in everyday business contexts (Khalil & Abdelli, 2022; Rust & Huang, 2014; Mende et al., 2019). In transitioning from mainly human-centric work to technology-infused working life, our model deepens the understanding of employees willing to collaborate with newly developing artificial intelligence at work. Our findings add to the developing research field and offer important theoretical and managerial insights on increasing employees' willingness to cooperate with technology and ensuring employees' wellbeing.

This research defines AI-based services, which builds on existing literature on service technology and extends the scope of the concept (Jörling et al., 2019; Mende et al., 2019; van Doorn, 2017). With this scope, this chapter can cover the reality of employees confronted with technology in their workplace, which individually shapes their willingness to collaborate.

Employees' perceived risk about AI-based services is identified as a central driver in the qualitative study of this distinctive decision of employees on how to collaborate with technology. Furthermore, it is additionally outlined that risk is influenced by perceived control over technology, technology characteristics, employee characteristics, and job characteristics.

Based on these findings and a thorough literature review, this research proposes the conceptual framework of employees' willingness to cooperate with AI-based services and its impact on employees' wellbeing. This study presents a first venture to understand how a positive co-existence of intelligent technologies and humans can be created in working contexts. With a special focus on perceived employee drawbacks, this article details aspects that may hinder the willingness to cooperate. Refusing or holding up the technology integration process in the workplace has crucial implications for enterprises. AI-based service technologies are objectively more efficient, leading to significant cost savings, fewer errors, and lower staffing levels (Bughin et al., 2018; Mogaji et al., 2020). Thus, companies must develop strategies to reduce employees' perceived risk along our proposed dimensions of technology characteristics, employee, job demands and resources, and control. All these efforts also positively influence employees' wellbeing in the workplace. Transitioning into a more technological world, our model can help managers address employees' concerns and implement AI-based services into their everyday work life (Abdulquadri et al., 2021). This chapter recommends that managers act along three dimensions to ensure employee cooperation.

Design processes with employees: Firstly, companies need to involve their employees in processes leading to AI-based service implementation. Employees need to be gradually and carefully introduced to AI and be completely informed about the advantages and capabilities of AI-based services. This understanding helps to lower acceptance barriers. *If AI helps me do my job better, why wouldn't I try it?* [IP12]. Existing employees need to be taught about the functionalities and implications of new technologies to increase technology readiness and decrease perceived risks about the adaption of AI. In addition, employees should be allowed to determine the areas of deployment and tasks of the AI in their workplace. *Hmm, I can personally decide which tasks I give, which I let solve by AI, um, and which tasks I do myself* [IP3]. This gives them a sense of appreciation for their work and reduces the fear of being replaced by technology.

Give employees control over the technology: In order to avoid lower automation levels which consequently can decrease the benefits of new technologies, managers can give employees the ability to interrupt or slightly influence procedures to increase cooperation intentions (Dietvorst et al.,

2018; Jörling et al., 2019). Quotations like this prove employees' pursuit of control. *You need control over the AI and must be able to shut it off anytime you like* [IP1]. It is also expected that job requirements will change (Segal, 2018; Wirtz et al., 2018. Over time employees might drift into the role of supervisors anyhow, so they regain control over technology again.

Select the right employees: Finally, selecting employees who support and enjoy working with advanced service technologies can also influence the willingness to cooperate. Based on the conceptual model, three central employee characteristics (technology readiness, social power, and need for control) are identified that managers should consider when choosing new staff members. Regarding social power, it seems crucial to implement rules and regulations regarding the collaboration between employees and AI-based services. Furthermore, since many interview partners have a clear idea about future collaboration in mind, *Humans will always be in the superior role, so employees have to make the final decision* [30]. An explicit assignment of roles is important, in which responsibilities, authority, and decision-making power are clarified and regulated to avoid misunderstandings during AI and human collaboration.

References

Abdulquadri, A., Kieu, T. A., & Nguyen, N. P. (2021). Digital transformation in financial services provision: Perspective to the adoption of chatbot. *Journal of Enterprising Communities: People and Places in the Global Economy, 15*(2), 258–281.

Bakker, A. B., Demerouti, E., & Verbeke, W. (2004). Using the job demands-resources model to predict burnout and performance. *European Journal of Work and Organizational Psychology, 43*(1), 83–104. https://doi.org/10.1002/hrm.84

Bateson, J. E. G. (1985). Self-service consumer-an exploratory-study. *Journal of Retailing, 61*(3), 49–76.

Bazeley, P., & Jackson, K. (Eds.). (2013). *Qualitative data analysis with NVivo*. Sage Publications Limited.

Beer, J. M., Fisk, A. D., & Rogers, W. A. (2014). Toward a framework for levels of robot autonomy in human-robot interaction. *Journal of Human-Robot Interaction, 3*(2), 74. https://doi.org/10.5898/JHRI.3.2

Biocca, Frank and Harms, C. (2002). Defining and measuring social presence: Contribution to the networked minds theory and measure. In F. R. Gouveia & F. Biocca (Eds.), *Proceedings of the 5th international workshop on presence* (pp. 7–36).

Birdi, K., Clegg, C., Patterson, M., Robinson, A., Stride, C. B., Wall, T. D., & Wood, S. J. (2008). The impact of human resource and operational management practices on company productivity: A longitudinal study. *Personnel Psychology, 61*(3), 467–501.

Boos, A., Herzog, O., Reinhardt, J., Bengler, K., & Zimmermann, M. (2022). A compliance–reactance framework for evaluating human-robot interaction. *Frontiers in Robotics and AI, 42*(3–4), 167–175. https://doi.org/10.1016/S0921-8890(02)00373-1

Breazeal, C. (2003). Toward sociable robots. *Robotics and Autonomous Systems, 42*(3–4), 167–175.

Bughin, J., Hazan, E., Lund, S., Dahlström, P., Wiesinger, A., & Subramaniam, A. (2018). *Skill shift: Automation and the future of the workforce*. McKinsey & Company.

Chau, A. W., & Phillips, J. G. (1995). Effects of perceived control upon wagering and attributions in computer blackjack. *Journal of General Psychology, 122*(3), 253–269.

Collier, J. E., & Sherrell, D. L. (2010). Examining the influence of control and convenience in a self-service setting. *Journal of the Academy of Marketing Science, 38*(4), 490–509. https://doi.org/10.1007/s11747-009-0179-4

Darwin, C. (1859). *On the origin of species*. John Murray.

Demerouti, E., Bakker, A. B., Nachreiner, F., & Schaufeli, W. B. (2001). The job demands-resources model of burnout. *Journal of Applied Psychology, 86*(3), 499.

Dietvorst, B. J., Simmons, J. P., & Massey, C. (2018). Overcoming algorithm aversion: People will use imperfect algorithms if they can (even slightly) modify them. *Management Science, 64*(3), 1155–1170.

Epley, N., Waytz, A., & Cacioppo, J. T. (2007). On seeing human: A three-factor theory of anthropomorphism. *Psychological Review, 114*(4), 864–886. https://doi.org/10.1037/0033-295X.114.4.864

Featherman, M. S., & Pavlou, P. A., 2003. Predicting e-services adoption: a perceived risk facets perspective. *International journal of human-computer studies, 59*(4), 451–474.

Frank, M. R., Autor, D., Bessen, J. E., Brynjolfsson, E., Cebrian, M., Deming, D. J., et al. (2019). Toward understanding the impact of artificial intelligence on labor. *Proceedings of the National Academy of Sciences, 116*(14), 6531–6539.

Fraune, M. R., Oisted, B. C., Sembrowski, C. E., Gates, K. A., Krupp, M. M., & Šabanović, S. (2020). Effects of robot-human versus robot-robot behavior and entitativity on anthropomorphism and willingness to interact. *Computers in Human Behavior, 105*, 106220. https://doi.org/10.1016/j.chb.2019.106220

Frewer, L. J., Shepherd, R., & Sparks, P. (1994). The interrelationship between perceived knowledge, control and risk associated with a range of food-related hazards targeted at the individual, other people and society. *Journal of Food Safety, 14*(1), 19–40. https://doi.org/10.1111/j.1745-4565.1994.tb00581.x

Hakanen, J. J., Schaufeli, W. B., & Ahola, K. (2008). The job demands-resources model: A three-year cross-lagged study of burnout, depression, commitment, and work engagement. *Work & Stress, 22*(3), 224–241. https://doi.org/10.1080/02678370802379432

Hinds, P. J., Roberts, T. L., & Jones, H. (2004). Whose job is it anyway ? A study of human–robot o t o y p o o t. *Human-Computer Interaction, 19*, 151–181. https://doi.org/10.1561/1100000005

Horswill, M. S., & McKenna, F. P. (1999). The effect of perceived control on risk taking. *Journal of Applied Social Psychology, 29*(2), 377–391. https://doi.org/10.1111/j.1559-1816.1999.tb01392.x

Huang, M.-H., & Rust, R. T. (2018). Artificial intelligence in service. *Journal of Service Research, 21*(2), 155–172.

Inesi, M. E., et al. (2011). Power and choice: Their dynamic interplay in quenching the thirst for personal control. *Psychological Science, 22*(8), 1042–1048. https://doi.org/10.1177/0956797611413936

Jones, G. R., & George, J. M. (1998). The experience and evolution of trust: Implications for cooperation and teamwork. *Academy of Management Review, 23*(3), 531–546. https://doi.org/10.5465/AMR.1998.926625

Jörling, M., Böhm, R., & Paluch, S. (2019). Service robots: Drivers of perceived responsibility for service outcomes. *Journal of Service Research, 22*(4), 404–420.

Khalil, A., & Abdelli, M. E. A. (2022). Do digital technologies influence the relationship between the COVID-19 crisis and SMEs' resilience in developing countries? *Journal of Open Innovation: Technology, Market, and Complexity, 8*(2), 100–109.

Kim, S., & McGill, A. L. (2011). Gaming with Mr Slot or gaming the slot machine? Power, anthropomorphism, and risk perception. *Journal of Consumer Research, 38*(1), 94–107. https://doi.org/10.1086/658148

Klein, W. M., & Kunda, Z. (1994). Exaggerated self-assessments and the preference for controllable risks. *Organizational Behavior and Human Decision Processes, 59*(3), 410–427. https://doi.org/10.1006/OBHD.1994.1067

Kock, N. (2004). The psychobiological model: Towards a new theory of computer-mediated communication based on Darwinian evolution. *Organization Science, 15*(3), 327–348. https://doi.org/10.1287/orsc.1040.0071

Krueger, R. A. (2014). *Focus groups: A practical guide for applied research.* Sage Publications.

Kunz, W., et al. (2018). Future service technologies: Business models, analytics, and experience. *Journal of Services Marketing, 33/4*, 502–506.

Langer, E. J., & Rodin, J. (1976). The effects of choice and enhanced personal responsibility for the aged: A field experiment in an institutional setting. *Journal of Personality and Social Psychology, 34*(2), 191.

Larivière, B., Bowen, D., Andreassen, T. W., Kunz, W., Sirianni, N. J., Voss, C., et al. (2017). "Service encounter 2.0": An investigation into the roles of technology, employees and customers. *Journal of Business Research, 79*, 238–246.

Li, X., et al. (2008). Why do we trust new technology? A study of initial trust formation with organizational information systems. *The Journal of Strategic Information Systems, 17*(1), 39–71. https://doi.org/10.1016/j.jsis.2008.01.001

Lin, Y.-T., Doong, H.-S., & Eisingerich, A. B. (2021). Avatar Design of Virtual Salespeople: Mitigation of recommendation conflicts. *Journal of Service Research, 24*(1), 141–159. https://doi.org/10.1177/1094670520964872

Makridis, C. A., & Mishra, S. (2022). Artificial intelligence as a service, economic growth, and wellbeing. *Journal of Service Research, 25*(4), 505–520. https://doi.org/10.1177/10946705221120218

McKenna, F. P. (1993). It won't happen to me: Unrealistic optimism or illusion of control? *British Journal of Psychology, 84*(1), 39–50. https://doi.org/10.1111/j.2044-8295.1993.tb02461.x

McLeay, F., Osburg, V. S., Yoganathan, V., & Patterson, A. (2021). Replaced by a robot: Service implications in the age of the machine. *Journal of Service Research, 24*(1), 104–121. https://doi.org/10.1177/1094670520933354

Mende, M., Scott, M. L., van Doorn, J., Grewal, D., & Shanks, I. (2019). Service robots rising: How humanoid robots influence service experiences and elicit compensatory consumer responses. *Journal of Marketing Research, 56*(4), 535–556.

Mogaji, E., & Nguyen, P. N. (2022). Managers' understanding of artificial intelligence in relation to marketing financial services: Insights from a cross-country study. *International Journal of Bank Marketing, 40*(6), 1272–1298.

Mogaji, E., Soetan, T., & Kieu, T. (2020). The implications of artificial intelligence on the digital marketing of financial services to vulnerable customers. *Australasian Marketing Journal., 29*(3), 235–242.

Montoya-Weiss, M. M., Voss, G. B., & Grewal, D. (2003). Determinants of Online Channel use and overall satisfaction with a relational, multichannel service provider. *Journal of the Academy of Marketing Science, 31*(4), 448–458. https://doi.org/10.1177/0092070303254408

Nass, C., Steuer, J., & Tauber, E. R. (1994, April). Computers are social actors. In *Proceedings of the SIGCHI conference on Human factors in computing systems* (pp. 72–78).

Paluch, S., Tuzovic, S., Holz, H. F., Kies, A., & Jörling, M. (2022). "My colleague is a robot" – Exploring frontline employees' willingness to work with collaborative service robots. *Journal of Service Management, 33*(2), 363–388. https://doi.org/10.1108/JOSM-11-2020-0406

Paluch, S., & Wunderlich, N. V. (2016). Contrasting risk perceptions of technology-based service innovations in inter-organizational settings. *Journal of Business Research, 69*(7), 2424–2431. https://doi.org/10.1016/j.jbusres.2016.01.012

Patton, M. Q. (2015). *Qualitative research & evaluation methods*. SAGE Publications.

Pavlou, P. A. (2003). Consumer acceptance of electronic commerce: Integrating trust and risk with the technology acceptance model. *International Journal of Electronic Commerce, 7*(3), 101–134. https://doi.org/10.1080/10864415.2003.11044275

Pitardi, V., Wirtz, J., Paluch, S., & Kunz, W. H. (2021). Service robots, agency and embarrassing service encounters. *Journal of Service Management, 9*(6), 101–134.

Prahl, A., & van Swol, L. (2017). Understanding algorithm aversion: When is advice from automation discounted? *Journal of Forecasting, 36*(6), 691–702. https://doi.org/10.1002/for.2464

Rahwan, I., Cebrian, M., Obradovich, N., Bongard, J., Bonnefon, J. F., Breazeal, C., et al. (2019). Machine behaviour. *Nature, 568*(7753), 477–486.

Rijsdijk, S., & Hultink, E. J. (2009). How today's consumers perceive tomorrow's smart products. *Journal of Product Innovation Management, 26*, 24–42.

Rijsdijk, S. A., & Hultink, E. J. (2003). "Honey, have you seen our hamster?" consumer evaluations of autonomous domestic products. *Journal of Product Innovation Management, 20*(3), 204–216. https://doi.org/10.1111/1540-5885.2003003

Ring, P. S., & van de Ven, A. H. (1994). Developmental processes of cooperative interorganizational relationships. *The Academy of Management Review, 19*(1), 90. https://doi.org/10.2307/258836

Rust, R. T., & Huang, M. H. (2014). The service revolution and the transformation of marketing science. *Marketing Science, 33*(2), 206–221.

Schaufeli, W. B., & Bakker, A. B. (2004). Job demands, job resources, and their relationship with burnout and engagement: A multi-sample study. *Journal of Organizational Behavior, 25*(3), 293–315. https://doi.org/10.1002/job.248

Schepers, J., Belanche, D., Casaló, L. V., & Flavián, C. (2022). How smart should a service robot be? *Journal of Service Research, 25*(4), 565–582. https://doi.org/10.1177/10946705221107704

Segal, M. (2018). How automation is changing work. *Nature, 563*, 132–135. https://doi.org/10.1038/d41586-018-07501-y

Sheridan, T. B., & Verplank, W. L. (1978). *Human and computer control of undersea teleoperators*. Man-Machine Systems Laboratory, Department of Mechanical Engineering, MIT.

Smith, K. G., Carroll, S. J., & Ashford, S. J. (1995). Intra- and interorganizational cooperation: Toward a research agenda. *Academy of Management Journal, 38*(1), 7–23. https://doi.org/10.2307/256726

Strauss, A., & Corbin, J. (1998). *Basics of qualitative research techniques*. Sage Publications.

Thrun, S. (2004). Towards a framework for human-robot interaction. *Human-Computer Interaction, 19*(1), 9–24.

van Doorn, J. (2017). Domo Arigato Mr Roboto: Emergence of automated social presence in organizational frontlines and customers' service experiences. *Journal of Service Research, 20*(1), 43–58. https://doi.org/10.1177/1094670516679272

Weinstein, N. D. (1980). Unrealistic optimism about future life events. *Journal of Personality and Social Psychology., 39*(5), 806–820.

White, R. W. (1959). Motivation reconsidered: The concept of competence. *Psychological Review, 66*, 297–333. https://doi.org/10.1037/h0040934

Wirtz, J., Kunz, W. H., Hartley, N., & Tarbit, J. (2022). Corporate digital responsibility in service firms and their ecosystems. *Journal of Service Research, 26*, 173. https://doi.org/10.1177/10946705221130467

Wirtz, J., Patterson, P. G., Kunz, W. H., Gruber, T., Lu, V. N., Paluch, S., & Martins, A. (2018). Brave new world: Service robots in the frontline. *Journal of Service Management, 20*(1), 9–24.

Wolf, I. (2015). The interaction between humans and autonomous agents. In M. Maurer et al. (Eds.), *Autonomous driving: Technical, legal and social aspects* (1st ed., pp. 103–124). Springer Open. https://doi.org/10.1007/978-3-662-48847-8_6

You, S., & Robert, L. (2018). Emotional attachment, performance, and viability in teams collaborating with embodied physical action (EPA) robots. *Journal of the Association for Information Systems, 19*(5), 377–407. https://doi.org/10.17705/1jais.00496

Index

© The Author(s), under exclusive license to Springer Nature Switzerland AG 2023
J. N. Sheth et al. (eds.), *Artificial Intelligence in Customer Service*,
https://doi.org/10.1007/978-3-031-33898-4

The manufacturer's authorised representative in the EU is Springer Nature Customer Service Centre GmbH, Europaplatz 3, 69115 Heidelberg, Germany. If you have any concerns regarding our products, please contact ProductSafety@springernature.com

Printed and bound by CPI Group (UK) Ltd, Croydon, CR0 4YY
28/09/2026
02224072-0014